Approximate Solution Methods in
Engineering Mechanics

Approximate Solution Methods in Engineering Mechanics

Second Edition

Arthur P. Boresi

Professor Emeritus, Department of Civil and Architectural Engineering, University of Wyoming, and Professer Emeritus, Department of Theoretical and Applied Mechanics, University of Illinois, Urbana-Champaign

Ken P. Chong

Director of Mechanics and Materials, National Science Foundation, and Guest Researcher, National Institute of Standards and Technology

Sunil Saigal

Professor of Civil Engineering, Carnegie Mellon University

John Wiley & Sons, Inc.

This book is printed on acid-free paper. ⊗

Copyright © 2003 by John Wiley & Sons, Inc. All rights reserved

Published by John Wiley & Sons, Inc., Hoboken, New Jersey
Published simultaneously in Canada

For general information on our other products and services or for technical support, please contact our Customer Care Department within the United States at (800) 762-2974, outside the United States at (317) 572-3993 or fax (317) 572-4002.

Wiley also publishes its books in a variety of electronic formats. Some content that appears in print may not be available in electronic books. For more information about Wiley products, visit our web site at www.wiley.com.

Library of Congress Cataloging-in-Publication Data:

Boresi, Arthur P. (Arthur Peter), 1924–
    Approximate solution methods in engineering mechanics / Arthur P. Boresi, Ken P. Chong, Sunil Saigal—2nd ed.
        p. cm.
    Includes bibliographical references and index.
    ISBN 0-471-40242-7
        1. Engineering mathematics—Data processing.   2. Approximation theory—Data processing.   3. Numerical analysis—Data processing.   I. Chong, K. P. (Ken Pin), 1942–
II. Saigal, Sunil.   III. Title.

TA335 .B67 2003
620.1—dc21
                                                                    2002027209

Printed in the United States of America

10 9 8 7 6 5 4 3 2 1

Contents

Preface

The widespread use of digital computers has had a profound effect in engineering and science. On the one hand, it has resulted in many benefits. On the other hand, because of inadequate training and experience of the user, it has often led to the "garbage in, garbage out" and "black box" syndromes. For example, with computers and appropriate software, we can model and analyze complex physical systems and problems. However, efficient and accurate use of numerical results obtained from computer programs requires considerable background and advanced working knowledge to avoid blunders and the blind acceptance of computer results. In this book we attempt to provide some of the background and knowledge necessary to avoid these pitfalls. In particular, we consider several of the most commonly used approximate methods employed in the solution of physical problems.

A realistic and successful solution of an engineering problem usually begins with an accurate physical model of the problem and a proper understanding of the assumptions employed. In turn, this physical model is transformed into a mathematical model or problem. The solution of the mathematical problem is usually obtained by numerical methods that by definition are approximate. Except for relatively simple cases, the successful use of numerical methods depends heavily upon the use of digital computers, ranging in power from microcomputers to supercomputers, depending on the complexity of the problem.

Powerful approximation (numerical) methods are presented in this book. These methods include:

- Weighted residuals methods
- Finite difference methods
- Finite element methods
- Finite strip/layer/prism methods
- Boundary element methods
- Meshless methods

The mathematical formulation of these methods is perfectly general. However, in this book, the applications deal mainly with the field of solid mechanics. An extensive list of references is provided.

Practicing engineers and scientists should find this book very readable and useful. Students may use the book both as a reference and as a text. This book is a sequel to an earlier book by the first two authors (*Elasticity in Engineering Mechanics,* published by Wiley, New York, 2000).

Arthur P. Boresi, Ken P. Chong, and Sunil Saigal

1

The Role of Approximate Solution Methods in Engineering

1.1 INTRODUCTION

The solution of an elasticity problem generally requires the description of the response of a material body (computer chips, machine part, structural element, or mechanical system) to a given excitation (such as force). In an engineering sense, this description is usually required in numerical form, the objective being to assure the designer or engineer that the response of the system will not violate design requirements. These requirements may include the consideration of deterministic and probabilistic concepts (Thoft-Christensen and Baker, 1982; Wen, 1984; Yao, 1985). In a broad sense the numerical results are predictions as to whether the system will perform as desired. The solution to the elasticity problem may be obtained by a direct numerical process (numerical stress analysis) or in the form of a general solution (which ordinarily requires further numerical evaluation).

The successful solution of a complex engineering problem begins with an accurate physical model of the problem. In turn, this physical model is transformed into a mathematical model. The solution of the mathematical model is usually obtained by numerical methods that are by definition approximate. The success of these numerical techniques rest in turn upon high-speed digital computers.

The finite difference method (FDM) (Mitchell and Griffiths, 1980; Tannehill et al., 1997) and the finite element method (FEM) (Bathe, 1995; Hughes, 1987; Cook et al., 1989; Zienkiewicz and Taylor, 1989) are widely used numerical techniques. These methods are classified as *domain methods,* in that the engineering system is analyzed either in terms of discretized finite grids (FDMs) or finite elements (FEMs) throughout the entire region of the system (body). Another method that has emerged as a powerful tool is the boundary element method (BEM) (Rizzo, 1967; Brebbia, 1978; Brebbia and Connor, 1989; Cruse, 1988). In certain problems, this method has some distinct advantages over FDM and FEM, for several reasons. In particular, a discretization of *only* the boundary of the domain of interest is necessary for BEM— hence the name *boundary element method.*

All three of the methods noted above, and a variety of other specialized techniques, provide powerful means of treating complex boundary value problems of engineering. In a particular case, depending on requirements, one of these methods may be more efficient than the others in generating a solution. For example, it may be more advantageous to use BEM for certain classes of linear problems characterized by infinite or semi-infinite domains, stress concentrations, three-dimensional structural effects, and so on (Beskos, 1989; Brebbia and Connor, 1989).

A fourth method, the finite strip method (FSM) [and the associated finite layer method (FLM) and finite prism method (FPM)] (Cheung and Tham, 1997) falls somewhere between domain methods (FDM and FEM) and boundary methods (BEM), in that it reduces the dimensions of the problem before discretization of the domain. Other hybrid methods that combine advantages of several formulations have been proposed. For example, Golley and Grice (1989), Golley et al. (1987), and Petrolito et al., (1989) combined FEM and FSM to study plate bending problems.

In this book we consider finite difference methods (Chapter 3), finite element methods (Chapter 4), finite strip (finite layer and finite prism) methods (Chapter 5), boundary elements (Chapter 6), and meshless methods (Chapter 7). In Chapter 2 we consider the fundamentals of approximate methods of analysis, including boundary solutions in terms of weighted residual methods (WRM). The various approximation techniques (FDM, FEM, FSM, BEM, etc.) may be represented as special cases of weighted residual formulations. They can be studied using the concepts of approximation and weighting that are fundamental to WRM and that are popular with engineers and mathematicians.

1.2 FIELDS OF APPLICATION

Approximation methods (Moin, 2001) are employed generally in all fields of engineering, mathematics, and science. For example, textbooks have been written dealing with applications in specialized subjects such as elasticity, plasticity, porous media flow, structural mechanics, fluid mechanics, aerodynamics, and so on (Boresi and Chong, 2000; Lubliner, 1990; Tannehill et al., 1997; Holzer, 1985). In this book we are concerned mainly with formulations of WRM, FDM, FEM, and FSM and with recent developments in computer implementations of these methods (Nelson, 1989). Selected references are given, with an emphasis on engineering applications in the solid mechanics area.

1.3 FUTURE PROGRESS AND TRENDS

Papers on the applications of the theory of elasticity to engineering problems form a significant part of the technical literature in solid mechanics (Dvorak, 1999; Boresi and Chong, 2000; Chong and Davis, 1999). Many of the solutions presented in current papers employ numerical methods and require the use of high-speed digital computers. This trend is expected to continue into the foreseeable future, particularly with the widespread use of microcomputers and workstations as well as the increased availability of supercomputers (Londer, 1985; Fosdick, 1996). For example, finite element methods have been applied to a wide range of problems, such as plane problems, problems of plates and shells, and general three-dimensional problems, including linear and nonlinear behavior, and isotropic and anisotropic materials. Furthermore, through the use of computers, engineers have been able to consider the optimization of large engineering systems (Atrek et al., 1984; Zienkiewicz and Taylor, 1989; Kirsch, 1993), such as the Space Shuttle. In addition, computers have played a powerful role in the fields of computer-aided design (CAD) and computer-aided manufacturing (CAM) (Ellis and Semenkov, 1983), as well as in virtual testing and model-based simulation (Fosdick, 1996; Chong et al., 2002). Recent advances in computation, in conjunction with synergistic combination with databased models (Garboczi et al., 2000), make virtual testing a valuable tool in designer materials with optimized properties.

The finite element method has limitations in solving problems with large deformations, crashes, fracture propagation, penetration, and other moving boundary problems. Mesh-free and particle methods, including smoothed particle hydrodynamics methods, mesh-free Galerkin methods, and molecular dynamics methods (Li and Liu, 2002), are especially useful for these classes of problems. Recently, hypersingular residuals in the mesh-based boundary element method have been extended to the meshless boundary node method with good potential for solving a wide range of problems efficiently (Chati et al., 2001).

In the past, engineers and material scientists have been involved extensively with the characterization of given materials. With the availability of advanced computing, along with new developments in material sciences, researchers can now characterize processes, design, and manufacture materials with desirable performance and properties. Using nanotechnology (Reed and Kirk, 1989; Timp, 1999; Siegel et al., 1999), engineers and scientists can build designer materials molecule by molecule.

Tremendous progress is being made in the measurement and application of microforces (see, e.g., Bowen et al., 1998; Saif and MacDonald, 1996). Advances have also been made in instrumentation, such as the AFM (atomic force microscope), SEM (scanning electron microscope), HRTEM (high-resolution transmission electron microscope), and SRES (surface roughness evolution spectroscope). Figure 1.1 summarizes the gauge length and strain resolution of various instruments, spanning scales from 10^{-9} to 10^{0} m. One of the challenges is to model short-term microscale material behavior through mesoscale and macroscale behavior into long-term structural systems performance (Fig.

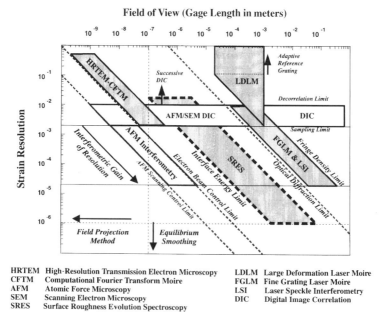

Figure 1.1 Map of deformation-measurement techniques. (Courtesy of K.-S. Kim, Nano & Micromechanics Laboratory, Brown University.)

1.2). Accelerated tests to simulate various environmental forces and impacts are needed. Supercomputers and/or workstations used in parallel are useful tools (1) to solve this multiscale and size-effect problem by taking into account the large number of variables and unknowns to project microbehavior into infrastructure systems performance and (2) to model or extrapolate short-term test results into long-term life-cycle behavior.

According to Eugene Wong, formerly of the National Science Foundation, the transcendent technologies of our time are:

- *Microelectronics:* Moore's law, the doubling of capabilities every 18 months for the past 30 years; unlimited scalability
- *Information technology:* confluence of computing and communications
- *Biotechnology:* molecular secrets of life.

Engineering mechanics and the associated numerical solution form the backbone of these transcendent technologies.

We stand at the threshold of an explosive expansion of computer applications in numerical solutions of problems in all areas of engineering (Tannehill et al., 1997; Nelson, 1989; Beskos, 1989). Major advances in finite element model technology have been outlined by Belytschko (1989; Belytschko et al., 2000). Numerical methods for solution of finite element systems are described by Wilson (1989). In addition, a broad range of topics on the application of modern computers to the solution of structural engineering problems is discussed in the proceedings of the 1989 Structures Congress (Nelson, 1989). These topics include artificial intelligence, parallel processing, optimization, knowledge-based systems, and computer-aided analysis and de-

MATERIALS		STRUCTURES		INFRASTRUCTURE	
nano level	micro level	meso level		macro level	systems level
(10^{-9}) m	(10^{-6})	(10^{-3})		(10^{+0})	(10^{+3}) m
Molecular Scale	*Micrometers*			*Meters*	*Up to Kilometer Scale*
nano mechanics	micro mechanics	meso mechanics		beams	bridge systems
self-assmbly	micro structures	interfacial structures		columns	lifelines
nanofabrication	smart materials	composites		plates	airplanes

Figure 1.2 Scales in materials and structures. (From Boresi and Chong, 2000.)

sign. The present status and possible future developments in boundary element methods in structural analysis are discussed by Banerjee and Mukherjee (1998), Cruse (1988), and Beskos (1989). Finally, it is anticipated that future work in meshless and hybrid methods will play a greater role in approximation methods (Chong et al., 2002; Chati et al., 2001; Golley and Grice, 1989; Costabel and Stephan, 1988).

REFERENCES

Atrek, E., Gallagher, R. H., Ragsdell, K. M., and Zienkiewicz, O. C. (eds.) 1984. *New Directions in Optimal Structural Design,* Wiley, New York.

Banerjee, P. K., and Mukherjee, S. 1998. *Developments in Boundary Element Methods,* E&FN Spon, London.

Bathe, K. J. 1995. *Finite Element Procedures,* Prentice Hall, Upper Saddle River, NJ.

Belytschko, T. 1989. Major advances in finite element model technology, in *Proc. Sessions Related to Computer Utilization at Structures Congress '89,* Nelson, J. K., Jr. (ed.), American Society of Civil Engineers, Reston, VA, pp. 11–20.

Belytschko, T., Liu, W. K., and Moran, B. 2000. *Nonlinear Finite Elements for Continua and Structures,* Wiley, New York.

Beskos, D. E. 1989. *Boundary Element Methods in Structural Analysis,* American Society of Civil Engineers, Reston, VA.

Boresi, A. P., and Chong, K. P. 2000. *Elasticity in Engineering Mechanics,* Wiley, New York.

Bowen, W. R., Hilal, N., Lovitt, R. W., and Wright, C. J. 1998. Direct measurement of the force of adhesion of a single biological cell using an atomic force microscope, *Colloids Surf. A: Physicochem. Eng. Aspects,* **136,** 231–234.

Brebbia, C. A. 1978. *The Boundary Element Method for Engineers,* Halsted Press, division of Wiley, New York.

Brebbia, C. A., and Connor, J. J. (eds.) 1989. *Advances in Boundary Elements,* Springer-Verlag, New York.

Chati, M. K., Paulino, G. H., and Mukherjee, S. 2001. The meshless standard and hypersingular boundary node methods, *Int. J. Numer. Methods Eng.,* **50,** 2233–2269.

Cheung, Y. K., and Tham, L. G. 1997. *Finite Strip Method,* CRC Press, Boca Raton, FL.

Chong, K. P., and Davis, D. C. 1999. Engineering mechanics and materials research in the information technology age, *Mechanics,* **28**(7/8).

Chong, K. P., Saigal, S., Thynell, S., and Morgan, H. S. (eds.) 2002. *Modeling and Simulation-Based Life-Cycle Engineering,* Spon, New York.

Cook, R. D., Malkus, D. S., and Plesha, M. E. 1989. *Concepts and Applications of Finite Element Analysis,* 3rd ed., Wiley, New York.

Costabel, M., and Stephan, E. P. 1988. Coupling of finite elements and boundary elements for transmission problems of elastic waves in R^3, in *Advanced Boundary Element Methods,* Cruse, T. A. (ed.), Springer-Verlag, New York, pp. 117–124.

Cruse, T. A. (ed.) 1988. *Advanced Boundary Element Methods,* Springer-Verlag, New York.

Dvorak, G. J. (ed.) 1999. Research trends in solid mechanics, *Int. J. Solids Struct.,* **37**(1/2): special issues.

Ellis, T. M. R., and Semenkov, O. I. (eds.) 1983. *Advances in CAD/CAM,* North Holland, Amsterdam.

Fosdick, L. D. (ed.) 1996. *An Introduction to High-Performance Scientific Computing,* MIT Press, Cambridge, MA.

Garboczi, E. J., Bentz, D. P., and Frohnsdorff, G. F. 2000. Knowledge-based systems and computational tools for concrete, *Concrete Int.,* **22**(12) 24–27.

Golley, B. W., and Grice, W. A. 1989. Prismatic folded plate analysis using finite strip-elements, in *Computer Methods in Applied Mechanics and Engineering,* Elsevier North-Holland, Amsterdam.

Golley, B. W., Grice, W. A., and Petrolito, J. 1987. Plate-bending analysis using finite strip-elements, *J. Struct. Eng.,* **113**(6).

Holzer, S. M. 1985. *Computer Analysis of Structures,* Elsevier Science, New York.

Kirsch, U. 1993. *Structural Optimization,* Springer-Verlag, New York.

Li, S. F., and Liu, W. K. 2002. Meshfree and particle methods and their applications, *Appl. Mech. Rev.,* Sept. 2002.

Londer, R. 1985. Access to supercomputers, *Mosaic,* **16**(3), 26–32.

Lubliner, J. 1990. *Plasticity Theory,* Macmillan, New York.

Moin, P. 2001. *Fundamentals of Engineering Numerical Analysis,* Cambridge University Press, Cambridge.

Nelson, J. K., Jr. (ed.) 1989. *Computer Utilization in Structural Engineering,* Proc. Structures Congress '89, American Society of Civil Engineers, Reston, VA.

Petrolito, J., Grice, W. A., and Golley, B. W. 1989. Finite strip-elements for thick plate analysis, *J. Struct. Eng.,* **115**(6).

Reed, M. A., and Kirk, W. P. (eds.) 1989. *Nanostructure Physics and Fabrication,* Academic Press, San Diego, CA.

Rizzo, F. J. 1967. An integral equation approach to boundary value problems of classical elastostatics, *Q. J. Appl. Math.,* **25,** 83–95.

Saif, M. T. A., and MacDonald, N. C. 1996. A milli-Newton micro loading device. *Sensors Actuators A,* **52,** 65–75.

Siegel, R. W., Hu, E., and Roco, M. C. (eds.) 1999. *WTEC Panel Report on Nanostructure Science and Technology,* Kluwer Academic Publishers, Norwell, MA.

Tannehill, J. C., Anderson, D. A., and Pletcher, R. H. 1997. *Computational Fluid Mechanics and Heat Transfer,* 2nd ed., Taylor & Francis, Philadelphia.

Thoft-Christensen, P., and Baker, M. J. 1982. *Structural Reliability Theory and Its Applications,* Springer-Verlag, Berlin.

Timp, G. (ed.) 1999. *Nanotechnology,* Springer-Verlag, New York.

Wen, Y.-K. (ed.). 1984. *Probabilistic Mechanics and Structural Reliability,* American Society of Civil Engineers, Reston, VA.

Wilson, E. L. 1989. Numerical methods for solution of finite element systems, in *Proc. Sessions Related to Computer Utilization at Structures Congress '89,* Nelson, J. K., Jr. (ed.), American Society of Civil Engineers, Reston, VA, pp. 21–30.

Yao, J. T. P. 1985. *Safety and Reliability of Existing Structures,* Pitman Advanced Publishing Program, Boston.

Zienkiewicz, O. C., and Taylor, R. L. 1989. *The Finite Element Method,* 4th ed., Vol. 1, *Basic Formulation and Linear Problems,* McGraw-Hill, New York.

BIBLIOGRAPHY

Brand, L. 1957. *Vector and Tensor Analysis,* Wiley, New York.

Chong, K. P., Dewey, B. R., and Pell, K. M. 1989. *University Programs in Computer-Aided Engineering, Design, and Manufacturing,* American Society of Civil Engineers, Reston, VA.

Danielson, D. A. 1997. *Vectors and Tensors in Engineering and Physics,* Addison-Wesley, Reading, MA.

Dym, C. L., and Shames, I. H. 1973. *Solid Mechanics: A Variation Approach,* McGraw-Hill, New York.

Edelen, D. G. B., and Kydoniefs, A. D. 1980. *An Introduction to Linear Algebra for Science and Engineering,* 2nd ed., Elsevier Science, New York.

Eisele, J. A., and Mason, R. M. 1970. *Applied Matrix and Tensor Analysis,* Wiley-Interscience, New York.

Gere, J. M., and Weaver, W. 1984. *Matrix Algebra for Engineers,* 2nd ed., Prindle, Weber, and Schmidt, Boston.

Hughes, T. J. R. 1987. *The Finite Element Method,* Prentice Hall, Upper Saddle River, NJ.

Jeffreys, H. 1987. *Cartesian Tensors,* Cambridge University Press, New York.

Kemmer, N. 1977. *Vector Analysis,* Cambridge University Press, Cambridge.

Luré, A. I. 1964. *Three-Dimensional Problems of the Theory of Elasticity,* Wiley-Interscience, New York.

Mitchell, A. R., and Griffiths, D. F. 1980. *The Finite Difference Method in Partial Differential Equations,* Wiley, New York.

Nash, W. A. 1993. *The Mathematics of Nonlinear Mechanics,* CRC Press, Boca Raton, FL.

Nemat-Nasser, S., and Hori, M. 1993. *Micromechanics,* North-Holland, Amsterdam.

Spanos, P. D. (ed.) 1999. *Computational Stochastic Mechanics,* A.A. Balkema, Rotterdam, The Netherlands.

Ting, T. C. T. 1996. *Anisotropic Elasticity,* Oxford University Press, New York.

2

Approximate Analysis and Weighted Residuals

2.1 INTRODUCTION

Exact analytical solutions to certain engineering boundary value problems exist (Boresi and Chong, 2000). However, in many cases, the boundary value problems of engineering cannot be solved exactly by currently available analytical methods. In such cases, approximate solutions are sought.

In some situations, we may be able to simplify the physical model suitably and obtain exact solutions to the modified problem. In other circumstances it may be advantageous to employ a direct attack on the equations through finite difference methods (Chapter 3) or by piecewise polynomial methods (such as finite element methods, Chapter 4). Alternatively, in some problems we may employ the classical method of separation of variables in conjunction with series methods (Chapter 5).

There exists a broad area of mathematics known as *approximation theory* (Shisha, 1968). The term is usually reserved for that branch of mathematics devoted to the approximation of general functions by means of simple functions. For example, in practice we may wish to approximate a real arbitrary function $f(x)$ by means of a polynomial $p(x)$ in some finite interval of space, say, $a \leqslant x \leqslant b$. The motivation for such an approximation is often one of simplification, particularly when the function $f(x)$ is too complicated to manipulate. Because the approximation, say $F(x)$, ordinarily differs from $f(x)$, questions arise immediately as to the manner in which we should proceed. As noted by Shisha, approximation theory considers such problems as the following:

1. What kinds of functions $F(x)$ should we consider for approximating $f(x)$? In other words, the question of function form or trial function for the approximation is considered.
2. How do we measure the accuracy or goodness of the approximation? To answer this question, we must establish some measure or norm of accuracy or goodness of approximation.
3. Of all possible forms or trial functions, is there one that approximates $f(x)$ well? How well? Is there, by some standard of measure (norm), one approximation that fits $f(x)$ best? That is, for a given norm, is there some best approximation of $f(x)$? If there is a best approximation, is it unique?
4. How can we get approximations to $f(x)$ in practice?

The problem of obtaining approximate solutions to initial value or boundary value problems differs from that of obtaining the best approximation of known functions in that the exact solution—call it $y(x)$—is unknown. In addition, the solution is required to satisfy a differential equation as well as initial values and/or boundary values. Hence, although we wish to approximate $y(x)$ by some trial function $Y(x)$ in the best possible sense, our problem is greatly complicated. Nevertheless, the ideas and concepts of general approximation theory can be employed in approximation solutions of initial value and/or boundary value problems of engineering.

In this chapter we discuss briefly concepts common to a broad class of approximation methods. In particular, we treat the *method of weighted residuals,* the most general of trial function methods (Crandall, 1956; Collatz, 1960; Finlayson, 1972). Depending on the norm employed, we may show that the method of weighted residuals leads to well-known approximation methods (e.g., the Galerkin method, collocation, least squares). In other words, the method of weighted residuals *unifies* many approximation methods that are currently in use (Finlayson and Scriven, 1966; Reddy, 1991; Hughes, 2000; Cook et al., 1989). For example, we will see that the Rayleigh–Ritz method (Langhaar, 1989) is a weighted residual method with a particular choice of weighting function.

In Chapter 3 we treat in detail approximate solutions of engineering problems by finite difference methods. In Chapter 4 we develop finite element methods and apply them to two- and three-dimensional problems. In Chapter 5 we formulate finite strip, finite layer, and finite prism methods for special applications. In Chapters 6 and 7 we present boundary element and meshless methods.

2.2   APPROXIMATION PROBLEM (TRIAL FUNCTIONS; NORMS OR MEASURES OF ERROR)

We are here concerned primarily with approximate methods of solving boundary value problems in engineering. In boundary value problems in engineering, we are generally faced with determining a solution to a differential equation (or a system of differential equations) in a region R. The solution is required to meet certain conditions on the boundary B of the region. In many cases the given equation or equations do not possess a known exact solution. Accordingly, unless we are able to obtain the unknown exact solution, we are forced to find an approximation of the exact solution. The form of this approximation is often cast in terms of a *trial solution* or *trial function* $F(x; A)$, which is assumed to be compatible with the exact solution $E(x)$, where x is defined in a compact subset (region) of space. For example, x may denote n real variables, or it may denote three spatial coordinates in the domain R; in one-dimensional problems x denotes a single real variable. The symbol A stands for a collection of parameters $a_1, a_2, a_3, \ldots, a_n$. Thus $A = A{:}(a_1, a_2, a_3, \ldots, a_n)$. Unfortunately, *there is no general scientific method of determining which of the unlimited number of approximating functions (trial functions or forms) ordinarily available will lead to the most efficient approximation of $E(x)$.* In practice, the choice of trial functions is often made on the basis of experience or intuition. For example, one may sense intuitively that a certain form of trial function (say, a polynomial or a Fourier series) may be suitable but not have any method at hand to actually determine the required approximation.

   The choice of a method of estimating the accuracy of the approximation (choice of norm) ordinarily is less important than that of trial function (choice of form). Generally speaking, if $F(x; A)$ is compatible with $E(x)$, almost any reasonable norm will lead to an efficient approximation of $E(x)$. However, if $F(x; A)$ is not compatible with $E(x)$, an efficient approximation will not ordinarily be attained, regardless of the norm employed. Of course, the choice of the norm may affect the complexity of estimating the accuracy of the approximation. In some cases this factor may dictate the choice of norm. In other words, once the form or family of trial functions for the approximation is selected, we wish to select the best approximation possible within the family. Then the norms on which this best approximation rests are essentially unlimited in number. To simplify the choice somewhat, we restrict the discussion that follows to two rather broadly employed methods of error measurement: the method of weighted residuals and the variational method. In either case, for linear boundary value problems, the method leads to consideration of the solution of a set of simultaneous linear algebraic equations.

In Section 2.3 we consider the method of weighted residuals as applied to ordinary differential equations, and in Section 2.4, to partial differential equations. In Section 2.5 we outline briefly the variation method and show its relation to the method of weighted residuals.

2.3 METHOD OF WEIGHTED RESIDUALS (ORDINARY DIFFERENTIAL EQUATIONS)

Preliminary Remarks

In certain simplified situations the boundary value problem may be reduced to one of ordinary differential equations in a single dependent variable. (See, for example the axially symmetric and spherically symmetric problems of elasticity; Boresi and Chong, 2000. See also the finite strip method of Section 5.1.) Furthermore, the method of separation of variables leads to ordinary differential equations (see Sections 6.9, 7.10, and 7.15, Boresi and Chong, 2000). Consequently, in this section we consider a method of evaluation of ordinary differential equations based on the method of weighted residuals. First, we outline the general approximation method for ordinary differential equations.

As noted in Section 2.2, an approximate solution to differential equations is often sought by assuming that the exact solution $E(x)$ may be approximated by an expression of the form $F(x; a_1, a_2, a_3, \ldots, a_n)$, where $a_1, a_2, a_3, \ldots, a_n$ are arbitrary parameters to be chosen to best fit the exact solution $E(x)$. This best fit is related to a particular measure (norm) of the approximation. In selecting the form $F(x; a_1, a_2, a_3, \ldots, a_n)$, we have certain options available. For example, let $R + B = D$ be the domain of the boundary value problem, where R is the interior of D bounded by surface B. Then we may choose F in one of the following ways (Collatz, 1960):

1. The differential equation is satisfied exactly in R, and the a_i are selected to make F fit the boundary conditions on B in some best

sense (norm). This method of selecting the a_i is called *the boundary method.*

2. The boundary conditions are satisfied exactly on B, and the a_i are selected so that F satisfies the differential equation in the interior R in some best sense (norm). This method is called *the interior method.*

3. The differential equation is not satisfied in R, nor are the boundary conditions satisfied on B. The a_i are chosen to satisfy the differential equation in R and the boundary conditions on B in some best sense. This method of determining the a_i is called a *boundary-interior method* or simply a *mixed method.*

In ordinary differential equations, interior methods are most often used, as even if we know the general solution to the differential equation (boundary method), we still must solve a set of n algebraic (linear or possibly nonlinear) equations in the a_i to satisfy the boundary conditions. In the boundary value problem of ordinary differential equations, the boundary conditions are usually specified at two points of the region, say, $x = 0$ and $x = L$. Thus, we speak of the two-point boundary value problem.

In boundary value problems of partial differential equations, both boundary and interior methods are used. However, in many cases boundary methods are preferred, as satisfaction of boundary conditions, as far as integration is concerned, requires the evaluation of integrals over the boundary B rather than evaluation of integrals through the interior region R, as do interior methods. When the differential equations and boundary conditions are very complicated, some simplification may be possible with the use of mixed methods.

Method of Weighted Residuals

The method of weighted residuals seeks to produce a best approximate solution to a differential equation (subject to boundary conditions) through the use of trial functions. This use is also a feature of variational methods (see Section 2.5, Chapter 4, and Langhaar, 1989). Special widely used cases of the method of weighted residuals include the Galerkin method, the method of collocation, and the method of least squares (Collatz, 1960; Botha and Pinder, 1983; Allen and Isaacson, 1997). The general approach may be outlined as follows: Consider the ordinary differential equation

$$G[y] - f(x) = 0 \qquad \text{for} \quad x_0 \leqslant x \leqslant x_1 \tag{2.1}$$

with boundary condition

$$B[y] = 0 \qquad \text{for} \quad x = x_0, \, x = x_1 \tag{2.1a}$$

where $f(x)$ is a known function of x, and G and B denote differential operators of x. In general, G and B may be nonlinear operators, for example,

$$G = C_1 \frac{d^2}{dx^2} + C_2 \left(\frac{d}{dx} \right)^2 + C_3$$

In linear boundary value problems, G and B are linear differential operators, for example,

$$G = C_1 \frac{d^2}{dx^2} + C_2 \frac{d}{dx} + C_3$$

The C's may be functions of x. The general solution of Eq. (2.1) is a function $y = y(x)$ that satisfies the differential equation.

In the method of weighted residuals, we may assume an approximate solution $\bar{y}$ of Eq. (2.1) of the form

$$y \approx \bar{y}(x: a_1, a_2, a_3, \ldots, a_n) = \sum_{i=1}^{n} a_i \phi_i(x) \tag{2.2}$$

where the a_i are undetermined parameters and the $\phi_i(x)$ are *trial functions* chosen so that $\bar{y}$ satisfies boundary conditions (the interior method). For example, if the boundary conditions are $y = 0$ for $x = a$ and for $x = b$, a possible choice of $\phi_i(x)$ is

$$\phi_i(x) = (x - a)(x - b)x^{i-1} \tag{2.2a}$$

In some problems a boundary condition may be nonlinear (in y). Then it may not be possible to select $\phi_i(x)$ such that $\bar{y}$ satisfies this boundary condition, and we are led to the mixed method of determining the a_i. The residual $r(x) = B(\bar{y})$ formed by substituting $\bar{y}$ and its derivatives for y and its derivatives into the boundary condition [see Eq. (2.1a)] is then not identically zero. We may obtain an equation in the a_i by liquidating the residual $r(x)$ in some way, for example, by arbitrarily setting it equal to zero. This equation must then be solved with the other equations by the method of weighted residuals.

For the *interior method,* the method of weighted residuals [Collatz (1960) calls it the orthogonality method] requires that the integral of the residual, $R(x) = G[\bar{y}] - f(x)$, appropriately weighted by functions $w_i(x)$, $i = 1, 2, 3, \ldots, n$, vanishes over the interior region $x = [0, L]$. [In the terminology of Collatz (1960), $R(x)$ is said to be orthogonal to $w_i(x)$ over the interior region $x = [0, L]$.] Thus,

$$\int_0^L w_i(x)R(x)\ dx = 0 \qquad i = 1, 2, 3, \ldots, n \qquad (2.3)$$

where $w_i(x)$ are n linearly independent functions. Theoretically, the $w_i(x)$ should be members of a complete set of functions. In practice, the $w_i(x)$ are often chosen to be the first n functions of a complete set, such as $\sin(2\pi x/L)$, $\sin(4\pi x/L)$, $\ldots$, $\sin(2\pi nx/L)$. The various trial function techniques are characterized by the choices for w_i (Finlayson, 1972), as discussed below.

The collocation method requires that the residual $R(x)$ vanish exactly at n points. Thus,

$$R(x_i) = 0 \qquad i = 1, 2, 3, \ldots, n \qquad (2.4)$$

Because Eq. (2.4) may be obtained from Eq. (2.3) with $w_i(x) = \delta(x - x_i)$, where $\delta(x - x_i)$ is the Dirac delta function defined by the conditions

$$\delta(x - x_i) = 0, \quad x \neq x_i \qquad \int_{-\infty}^{+\infty} \delta(x - x_i)\ dx = 1 \qquad (2.5)$$

the weighting function for the method of collocation is the Dirac delta function. Thus,

$$\int_0^L \delta(x - x_i)R(x)\ dx = R(x_i) = 0 \qquad i = 1, 2, 3, \ldots, n \quad (2.6)$$

and this weighting is equivalent to making the residual vanish at n

chosen points x_i. The collocation method has the mathematical advantage that the integration is trivial. However, because the forced solution is made to agree exactly at n selected points, it may fluctuate widely between points. For multivariable problems, Yang and Zhao (1985) addressed this difficulty by requiring residuals to vanish on selected *location lines.* Seemingly, this method has certain advantages over the point-location (collocation) method in multivariable problems.

The method of least squares requires that the integral of the residual squared be minimal. Thus, it requires that the integral

$$I = \int_0^L p(x)R^2(x)\ dx \tag{2.7}$$

be a minimum with respect to the a_i, where $p(x)$ is an arbitrary positive function. [Often, $p(x)$ is taken equal to 1.] Hence, the trial parameters a_i are determined from the condition

$$\frac{\partial I}{\partial a_i} = 0 \qquad i = 1, 2, 3, \ldots, n \tag{2.8}$$

Accordingly, in general the weighting functions in the method of least squares are

$$w_i(x) = p(x)\frac{\partial R(x)}{\partial a_i} \tag{2.9}$$

Unfortunately, in the method of least squares, the integrations involved in Eq. (2.3) are often very complicated.

In the *Galerkin method,* the most popular weighted method, the weighting functions $w_i(x)$ are taken to be the trial functions ϕ_i themselves. Hence, the residual is forced to be orthogonal to the trial functions $\phi_i(x)$. Thus, in the Galerkin method the trial parameters a_i are determined by the n equations

$$\int_0^L \phi_i(x)R(x)\ dx = 0 \qquad i = 1, 2, 3, \ldots, n \tag{2.10}$$

Accordingly, by the requirements of the method of weighted residuals, the trial functions $\phi_i(x)$ should be terms of a complete set of functions.

However, this condition, required for mathematical purposes, is often ignored in practice. Indeed, in practice at best the residual is made orthogonal to only the first few terms in a complete set.

In the *method of moments,* the residual is made orthogonal to the terms of a system of functions that need not be the same as the trial functions. Often, the method of moments is taken to be defined by the choice $w_i(x) = x^{i-1}$ regardless of the choice of $\phi_i(x)$. However, broadly speaking, the method of moments may be considered equivalent to the method of weighted residuals (or the method of orthogonalization). More narrowly, the method of moments is defined by the conditions

$$\int_0^L x^{i-1} R(x)\, dx = 0 \qquad i = 1, 2, 3, \ldots, n \qquad (2.11)$$

The *partition method* (or the *subdomain collocation method*) requires that the integral of the residual equals zero over m subintervals $s_i = [x_i, x_{i-1}]$, $i = 1, 2, 3, \ldots, m$ of $x = [0, L]$. Thus,

$$\int_{x_{i-1}}^{x_i} R(x)\, dx = 0 \qquad i = 1, 2, 3, \ldots, m \qquad (2.12)$$

Hence, the weighting functions $w_i(x)$ of Eq. (2.3) may be considered to be the unit step functions (see Chapter 4, where trial functions for finite element methods are discussed; see also Zienkiewicz and Morgan (1983))

$$w_i = \begin{cases} 1 & x \text{ in } s_i \\ 0 & x \text{ not in } s_i \end{cases} \qquad (2.13)$$

Boundary Methods

In the preceding discussion we considered primarily interior methods in which the trial functions are chosen such that $\bar{y}$ satisfies the boundary conditions but not the differential equation. For boundary methods we select trial functions so that the differential equation will be satisfied but not the boundary conditions. Then the procedure follows through as for interior methods, except that the averages over the interior [Eq. (2.3)] are replaced by appropriate equations on the boundary.

For discussions of mixed methods, refer to the comprehensive review of the method of weighted residuals given by Finlayson and Scriven

(1966); see also Collatz (1960, Chap. III), Finlayson (1972), and Cook et al. (1989).

Convergence Theorems

A few theorems have been proved concerning the convergence of methods of weighted residuals. These theorems pertain primarily to linear problems (Finlayson and Scriven, 1966; see also Bathe, 1995). When dealing with nonlinear problems, we must consider the possibility of the existence of more than one solution. The results obtained by successive approximations (with $n = 1, 2, 3, \ldots, N$) are often compared to justify convergence. Essential in this process is the choice of convergence tolerances (Bathe, 1995). If the tolerances are too coarse, inaccurate results may result. If the tolerances are too fine, excessive computational effort may be required, with little gain in accuracy. Also, a physical knowledge of the problem assists in judging whether the solution appears to be reasonable.

Example 2.1: Suspended Heavy Chain; Collocation Method, n = 1
A heavy chain is suspended from its endpoints located at equal heights at $x = 0$ and at $x = 1$. The deflection of the chain is approximated by solution of the differential equation $y'' + k(y')^2 + 1 = 0$, where k is the weight per unit length of chain divided by the tension in the chain at midspan. Primes denote derivatives relative to x. Assume an approximation $\bar{y}$ [Eq. (2.2)], with ϕ_i given by Eq. (2.2a).

(a) For $n = 1$, with $a = 0$, $b = 1$, $k = \frac{1}{2}$, $\bar{y}(0) = \bar{y}(1) = 0$, let $x_1 = \frac{1}{2}$ and determine a_1 by the collocation method [Eq. (2.4)].
(b) Compare the results with the exact solution

$$y = 2 \ln \frac{\cos(x/\sqrt{2} - 1/2\sqrt{2})}{\cos(1/2\sqrt{2})} \tag{2.14}$$

SOLUTION   Let

$$\bar{y} = \sum_{n=1}^{N} a_n \phi_n(x) \qquad \text{with} \quad \phi_n = (a - x)(b - x)x^{n-1}$$

(a) For $N = 1$, $a = 0$, $b = 1$, $k = \frac{1}{2}$, $\bar{y}(0) = \bar{y}(1) = 0$, we get $\phi_1(x) = x(x - 1)$. Hence, $\bar{y}(x) = a_1 x(x - 1)$, $\bar{y}'(x) = a_1(2x -$

1), $\bar{y}'' = 2a_1$, and $R(x) = \bar{y}'' + k(\bar{y}')^2 + 1 = 2a_1 + \frac{1}{2}a_1^2(2x - 1)^2 + 1$. For $x_1 = \frac{1}{2}$, $R(x_1) = 2a_1 + 1 = 0$. Therefore, $a_1 = -\frac{1}{2}$ and $\bar{y} = -(x^2 - x)/2$. Consider the values $x_1 = \frac{1}{2}$ and $x_2 = \frac{3}{4}$. For these values of x, $\bar{y}(\frac{1}{2}) = 0.125$ and $\bar{y}(\frac{3}{4}) = 0.09375$.

(b) By Eq. (2.14), $y(0) = y(1) = 0$, $y(\frac{1}{2}) = 0.1277$, and $y(\frac{3}{4}) = 0.09628$.

Example 2.2: Suspended Heavy Chain; Collocation Method, n = 2
Repeat Example 2.1 for $n = 2$ with $x_1 = \frac{1}{4}$, $x_2 = \frac{1}{2}$. Compare $\bar{y}(\frac{1}{2})$ with the results of Example 2.1. Compare $\bar{y}(\frac{1}{4})$ with $\bar{y}(\frac{3}{4})$.

SOLUTION   Let $n = 2$ in Example 2.1. Then by Eq. (2.2a), $\phi_2(x) = (x - a)(x - b)x$. Therefore, for $a = 0$, $b = 1$, $\phi_2(x) = x^3 - x^2$. Hence, by Eq. (2.2), with $\phi_1 = (x^2 - x)$ from Example 2.1,

$$\bar{y} = a_1\phi_1(x) + a_2\phi_2(x) = a_1(x^2 - x) + a_2(x^3 - x^2) \quad (2.15)$$

Therefore,

$$\bar{y}' = a_1(2x - 1) + a_2(3x^2 - 2x)$$
$$\bar{y}'' = 2a_1 + a_2(6x - 2)$$

and

$$R(x) = \bar{y}'' + \frac{1}{2}(\bar{y}')^2 + 1 = 2a_1 + a_2(6x - 2) + \frac{1}{2}[(2x - 1)^2a_1^2$$
$$+ 2a_1a_2(2x - 1)(3x^2 - 2x) + a_2^2(3x^2 - 2x)^2] + 1$$

Hence,

$$R\left(\frac{1}{4}\right) = 2a_1 - \frac{1}{2}a_2 + \frac{1}{8}a_1^2 + \frac{5}{32}a_1a_2 + \frac{25}{512}a_2^2 + 1 = 0$$
$$R\left(\frac{1}{2}\right) = 2a_1 + a_2 + \frac{1}{32}a_2^2 + 1 = 0$$
(2.16)

The solution of Eq. (2.16) is

$$a_1 = -0.5103 \qquad a_2 = 0.0206 \quad (2.17)$$

Equations (2.15) and (2.17) yield

$$\bar{y}(\tfrac{1}{2}) = 0.125 \qquad \text{[see Example 2.1(b)]}$$

and

$$\bar{y}(\tfrac{1}{4}) = 0.09472 \qquad \bar{y}(\tfrac{3}{4}) = 0.09278$$

Example 2.3: Suspended Heavy Chain; Galerkin Method Repeat Example 2.1 using the Galerkin method [Eq. (2.10)]. Compare the results with Example 2.1 for $\bar{y}(\tfrac{1}{2})$.

SOLUTION   From Example 2.1, with $n = 1$, $x = 0$, $x = 1$, $\phi_1(x) = x(x - 1)$, $\bar{y}(x) = a_1(x^2 - x)$, and $R(x) = 2a_1 + \tfrac{1}{2}a_1^2(4x^2 - 4x + 1) + 1$, we have by Eq. (2.3),

$$\int_0^1 (x^2 - x)[2a_1 + \tfrac{1}{2}a_1^2(4x^2 - 4x + 1) + 1]\, dx = 0$$

Integration yields $20a_1 + a_1^2 = -10$. The solution of this equation is $a_1 = -0.513$. Hence, $\bar{y}(x) = -0.513(x^2 - x)$. Thus, we obtain $\bar{y}(\tfrac{1}{2})_{\text{Galerkin}} = 0.128$. This result agrees well with that obtained by the exact solution, Example 2.1(b).

Example 2.4: Suspended Heavy Chain; Method of Moments Repeat Example 2.3 by the method of moments [Eq. (2.11)].

SOLUTION   As in Example 2.2, we have $R(x) = 2a_1 + \tfrac{1}{2}a_1^2(4x^2 - 4x + 1) + 1$. Hence, by Eq. (2.11), with $i = 1$, we have

$$\int_0^1 [2a_1 + \tfrac{1}{2}a_1^2(4x^2 - 4x + 1) + 1]\, dx = 0$$

Integration yields $12a_1 + a_1^2 = -6$. The solution to this equation is $a_1 = -0.5228$. Hence, $\bar{y}(x) = -0.5228(x^2 - x)$ and $\bar{y}(\tfrac{1}{2}) = 0.1307$, compared to $y(\tfrac{1}{2}) = 0.128$.

2.4 METHOD OF WEIGHTED RESIDUALS (PARTIAL DIFFERENTIAL EQUATIONS)

Many of the methods discussed in Section 2.3 for ordinary differential equations carry over to partial differential equations in a straightforward manner. However, the difficulties inherent in the approximate solutions of ordinary differential equations are magnified for partial differential equations. Indeed, even more than for ordinary differential equations, there is a need for further study of the existence and uniqueness of exact solutions. Perhaps more important to the stress analyst is the fact that the question of convergence of approximating trial functions is answered only for limited classes of problems (Zienkiewicz and Morgan, 1983; Allen and Isaacson, 1997). Unfortunately, calculations applied to partial differential equations may well lead to incorrect results. Finally, again most frustrating to the stress analyst, there is the fact that approximate methods may converge very nicely *but to values that are unrelated to the correct solution.* In early work on finite element methods, the latter phenomenon was observed frequently (see Chapter 4). It remains an important consideration in finite difference methods as well (see Chapter 3). Nevertheless, because the finite difference method is applicable to boundary value problems in general, it finds great favor. In particular, the approximate equations are easy to set up and even with a coarse mesh may give an approximate solution that is sufficiently accurate for the purposes of numerical stress analysis. One of the greatest disadvantages of the method is its slow rate of convergence, a difficulty when fine accuracy is required. [Collatz (1960) presents many examples of this problem in his Chapter V; see also our Chapter 3.]

In the following we briefly outline the method of weighted residuals as applied to boundary value problems of partial differential equations. The equations of elasticity are treated by difference methods in Chapter 3 and by finite element methods in Chapter 4. Because a number of important problems of elasticity reduce to the treatment of partial differential equations of second order, we restrict our attention to these types of equations. Treatment of higher-order equations follows by analogy.

For partial differential equations we seek to approximate the exact solution, say, $u(x_1, x_2, \ldots, x_n)$ by a trial function $\phi(x_1, x_2, \ldots, x_n; a_1, a_2, \ldots, a_p)$, where $x_1, x_2, \ldots, x_n$ denote independent variables and $a_1, a_2, \ldots, a_p$ are p arbitrary parameters. The trial function ϕ satisfies either the differential equation (boundary method) or the boundary conditions (interior method), whichever is more convenient. Substitution of ϕ into either the boundary conditions or the differential equations yields the residual error function $R(x_i, a_p)$. By choosing the parameters a_p so that the residual R is liquidated, say, in the sense of Eq. (2.3), we determine $\phi(x_i, a_p)$ to approximate $u(x_i, a_p)$ in the weighted residual sense.

Example 2.5: Torsion Problem by the Interior Method Consider the torsion problem defined by the equation (see Boresi and Chong, 2000, Sec. 7.3).

$$\nabla^2 u = -2 \qquad \text{in the interior of } D$$
$$u = 0 \qquad \text{on the boundary of } D \tag{2.18}$$

where for simplicity, region D is taken as the square $|x| \leq 1$, $|y| \leq 1$.

For the approximating function $\phi(x, y; a_0, a_1, \ldots, a_p)$, we may assume an expression that satisfies the differential equation (boundary method) or the boundary conditions (interior method). For example, in the interior method, to satisfy the boundary conditions we may choose a polynomial that meets any symmetry conditions that exist but whose coefficients are otherwise arbitrary. The coefficients are then selected to meet the boundary condition requirements. Thus, we take

$$\phi(x, y; a_0, a_1, \ldots, a_p) = a_0 + a_1(x^2 + y^2) + a_2 x^2 y^2$$
$$+ a_3(x^4 + y^4) + a_4(x^4 y^2 + x^2 y^4)$$
$$+ \cdots \tag{2.19}$$

Retaining only three terms in Eq. (2.19), as $u = 0$ on the boundary B, we find the one-parameter trial function

$$\phi(x, y; a) = a(1 - x^2 - y^2 + x^2 y^2) = a v_1(x, y) \tag{2.20}$$

where $v_1(x, y) = 1 - x^2 - y^2 + x^2 y^2$. Retaining five terms in Eq. (2.19) we obtain the two-parameter trial function

$$\phi(x, y; a, b) = a v_1(x, y) + b v_2(x, y) \tag{2.21}$$

where

$$v_1(x, y) = 1 - x^2 y^2 - x^4 - y^4 + x^2 y^2(x^2 + y^2)$$
$$v_2(x, y) = x^2 + y^2 - 2x^2 y^2 - x^4 - y^4 + x^2 y^2(x^2 + y^2)$$

Proceeding in this manner, we may include more parameters in the trial function ϕ for the interior method.

Alternatively, the differential equation [Eq. (2.18)] may be satisfied by the superposition of a particular integral (solution) and a series of harmonic functions that satisfy the homogeneous equation $\nabla^2 u = 0$ and also any symmetries of the problem (see Boresi and Chong, 2000, Sec. 7.5). Thus for a particular integral we take $-\frac{1}{2}(x^2 + y^2)$, and for the homogeneous equation we take the real parts of the complex variable $z = x + iy$ raised to the $4n$ power, where $n = 1, 2, 3, \ldots$. Then we have the $n + 1$ parameter trial function for the boundary method:

$$\phi(x, y; a_0, a_1, \ldots, a_n) = -\tfrac{1}{2}(x^2 + y^2) + a_0 + a_1(x^4 - 6x^2y^2 + y^4)$$

$$+ a_2(x^8 - 28x^6y^2 + 70x^4y^4 - 28x^2y^6 + y^8)$$

$$+ \cdots + a_n \, \mathrm{Re}(x + iy)^{4n} \qquad (2.22)$$

where Re denotes the real value. The method of weighted residuals may now be used to determine the parameters a_n. To determine the parameters a_n, we consider the collocation method and the Galerkin method in the following examples.

Collocation Method

First consider Eq. (2.20) to illustrate collocation as an *interior* method. Substitution into the differential equation (2.18) yields the residual

$$R(x, y; a) = \nabla^2 \phi + 2 = -2a(2 - x^2 - y^2) + 2 \qquad (2.23)$$

The coefficient a must be determined by setting the residual $R(x, y; a)$ equal to zero at some value of (x, y). The choice of the point (x, y) is rather arbitrary, although in the case of polynomials involving several parameters, $a_1, a_2, \ldots$, an attempt is usually made to select the collocation points (x_1, y_1), (x_2, y_2), $\ldots$, more or less uniformly spaced in the region D. Generally, there is no reliable procedure to assess the effect of the choice of collocation points. For the one-parameter example considered here, the parameter a is chosen so that $R(x, y; a)$ vanishes at one point. Thus, Eq. (2.23) yields the result

$$a = \frac{1}{2 - x^2 - y^2} \qquad (2.24)$$

for an arbitrary position of the single collocation point. Because for the interior of region D, $-1 < x < 1$ and $-1 < y < 1$, a may vary from the value $\frac{1}{2}$ (for $x = y = 0$) to a value as large as desired as $|x| \rightarrow 1$, $|y| \rightarrow 1$. Hence, the value of $\phi(0, 0; a)$ changes drastically with the choice of collocation point, a rather unsatisfactory result.

For the two-term approximation $\phi(x, y; a_1, a_2)$ of $u(x, y)$, we must consider two collocation points. Because of the symmetries that exist in the problem, this results in collocation at 16 points in the interior of region D. For example, if we take the points $x_1 = \frac{1}{2}$, $y_1 = \frac{1}{4}$ and $x_2 = \frac{3}{4}$, $y = \frac{1}{2}$ for collocation, by symmetry the points $(x = \pm\frac{1}{2}, y = \pm\frac{1}{4})$, $(x = \pm\frac{3}{4}, y = \pm\frac{1}{2})$, $(x = \pm\frac{1}{2}, y = \pm\frac{3}{4})$, and $(x = \pm\frac{1}{4}, y = \pm\frac{1}{2})$ are collocation points. Again the value of $\phi(0, 0; a_1, a_2)$ changes with collocation points but not widely as for the one-parameter approximation.

Galerkin Method

Again consider Eq. (2.20) to illustrate the Galerkin method (Cook et al., 1989) as an *interior* method. As in the collocation method, the residual is given by Eq. (2.23). Hence, because of the symmetry of the problem, the coefficient a is determined by the equation

$$\int_0^1 \int_0^1 v_1(\nabla^2\phi + 2) \, dx \, dy = 0 \qquad (2.25)$$

where v_1 is given by Eq. (2.20) and $(\nabla^2\phi + 2)$ is given by Eq. (2.23). Evaluation of the integral of Eq. (2.25) yields $a = 0.625$ and $\phi(0, 0; a) = 0.625$.

For the two-term approximation with (a, b), the Galerkin method requires that

$$\int_0^1 \int_0^1 v_i(\nabla^2\phi + 2) \, dx \, dy = 0 \qquad i = 1, 2 \qquad (2.26)$$

where $\phi = av_1 + bv_2$, and (v_1, v_2) are given by Eq. (2.21). Integration of Eq. (2.26) yields

$$1768a + 640b = 735$$
$$320a + 176b = 105 \qquad (2.27)$$

Hence, $a = 0.584386$, $b = -0.465930$, and $\phi(0, 0) = 0.584386$. Further refinement is possible by retaining more terms in the trial function representation [Eq. (2.19)].

2.5   VARIATION METHOD (RAYLEIGH–RITZ METHOD)

Generally speaking, the calculus of variations is a mathematical technique concerned with the determination of a function (say, y) that makes stationary a certain integral (say, J) of the function (Reddy, 1991). The integral or functional J takes on a particular numerical value for each function y. Hence, the functional J is said to be a function of the function y. For example, we may write for the one-dimensional case

$$J(y) = \int y(x) \, dx \tag{2.28}$$

Then for each function $y(x)$, $J(y)$ takes on a particular numerical value.

The fundamental problem of the calculus of variation is to determine a function y such that increments δy (called variations) of y result in second-order increments (or higher) in the functional J. Then J is said to be stationary. Symbolically, this means that when

$$y \rightarrow y + \delta y \tag{2.29}$$

then

$$J(y) \rightarrow J(y) + O(\delta y^2) \tag{2.30}$$

Thus, the term $\delta J(y)$ of first degree in y must vanish identically. The requirement that $J(y)$ be stationary, that is, $\delta J(y) = 0$, for arbitrary variations δy of y leads to an equation for y (the Euler equation; Langhaar, 1989). In physical problems the solution y of this equation is subject to certain boundary conditions. Then the variation δy of y must not violate these boundary conditions. For example, in the one-dimensional case, if $y(a) = C_1$ and $y(b) = C_2$, then $\delta y(a)$ and $\delta y(b)$ must be zero.

In the simplest variational problems, the integrand of the functional J does not contain derivatives higher than the first order in y. Thus for the case of one independent variable x,

$$J = \int_{x_0}^{x_1} F(x, y, y') \, dx \tag{2.31}$$

where $F(x, y, y')$ is a given function, y is a function of x, and the prime denotes derivative with respect to x. For J to be stationary for $y \rightarrow y + \delta y$, $\delta J = 0$. Thus, by Eqs. (2.29), (2.30), and (2.31),

$$\delta J = \int_{x_0}^{x_1} \delta F \, dx = 0 \tag{2.32}$$

where the first variation δF of F is given by

$$\delta F = \frac{\partial F}{\partial y} \delta y + \frac{\partial F}{\partial y'} \delta y' \tag{2.33}$$

where

$$\delta y' = \frac{d}{dx}(\delta y) \tag{2.34}$$

By Eqs. (2.32), (2.33), and (2.34), integration by parts yields

$$\delta J = \int_{x_0}^{x_1} \delta y \left(\frac{\partial F}{\partial y} - \frac{d}{dx} \frac{\partial F}{\partial y'} \right) dx + \frac{\partial F}{\partial y'} \delta y \bigg|_{x_0}^{x_1} \tag{2.35}$$

Since δy is arbitrary (except for possible requirements on the boundary $x = x_0$, $x = x_1$), for $\delta J = 0$, the integrand of Eq. 2.35 must vanish identically. Thus,

$$\frac{\partial F}{\partial y} - \frac{d}{dx} \frac{\partial F}{\partial y'} = 0 \tag{2.36}$$

Equation (2.36) is called the *Euler equation* of the functional J. For the one-dimensional case, it is an ordinary differential equation.

If the function $y(x)$ is required to satisfy boundary conditions at $x = x_0$, $x = x_1$, then

$$y(x_0) = C_1 \qquad y(x_1) = C_2 \tag{2.37}$$

where C_1, C_2 are constants. It follows that $\delta y(x_0) = \delta y(x_1) = 0$, and the terms outside the integral on Eq. (2.35) vanish identically. Alternatively, if y is free to take on arbitrary variations at $x = x_0$, $x = x_1$, for $\delta J = 0$, it is necessary that

$$\left.\frac{\partial F}{\partial y'}\right|_{x_0} = \left.\frac{\partial F}{\partial y'}\right|^{x_1} = 0 \tag{2.38}$$

These conditions enter naturally from Eq. (2.35) when y is not constrained by boundary conditions at x_0, x_1. Accordingly, Eqs. (2.38) are called the *natural boundary conditions* of the variation problem of seeking a function $y(x)$ that makes J stationary.

In some physical problems, such as stability problems, it is necessary to determine whether the stationary value of J corresponds to a maximum value, a minimum value, or neither (a *saddle-point value*). Then it is necessary to examine the higher-degree terms in Eq. (2.30). In other words, the nature of the second variation $\delta^2 J = \delta(\delta J)$ must be examined (Langhaar, 1989, Chap. 6).

More generally, if $F = F(x, y', y'', y''', \ldots)$, the Euler equation for F is

$$\frac{\partial F}{\partial y} - \frac{d}{dx}\frac{\partial F}{\partial y'} + \frac{d^2}{dx^2}\frac{\partial F}{\partial y''} - \frac{d^3}{dx^3}\frac{\partial F}{\partial y'''} + \cdots = 0 \tag{2.39}$$

If $F = F(x, y, z, y', z', y'', z'', \ldots)$, where y and z are functions of x, we obtain two Euler equations for F in the form

$$\frac{\partial F}{\partial y} - \frac{d}{dx}\frac{\partial F}{\partial y'} + \frac{d^2}{dx^2}\frac{\partial F}{\partial y''} - \frac{d^3}{dx^3}\frac{\partial F}{\partial y'''} + \cdots = 0$$
$$\frac{\partial F}{\partial z} - \frac{d}{dx}\frac{\partial F}{\partial z'} + \frac{d^2}{dx^2}\frac{\partial F}{\partial z''} - \frac{d^3}{dx^3}\frac{\partial F}{\partial z'''} + \cdots = 0 \tag{2.40}$$

If $F = F(x, y, w, w_x, w_y)$, where x and y are independent variables, $w = w(x, y)$, and subscripts x and y denote partial differentiation, the Euler equation is the partial differential equation

$$\frac{\partial F}{\partial w} - \frac{d}{dx}\frac{\partial F}{\partial w_x} - \frac{d}{dy}\frac{\partial F}{\partial w_y} = 0 \tag{2.41}$$

Generalization of Eq. (2.41) for several dependent variables and higher derivatives parallels the generalizations to Eqs. (2.39) and (2.40).

Approximation Techniques Based on Variational Methods

The approximation problem outlined in Section 2.2 may be attacked by variation methods. For example, let the Euler equation for the foundational J be of the form

$$H\phi - f = 0 \tag{2.42}$$

where H is a differential operator, ϕ is the function for which an exact solution is unknown, and f is a known function of the independent variables. Accordingly, let ϕ_i be a set of N trial functions for ϕ. Then, by Section 2.2, we seek N parameters a_i such that

$$\overline{\phi} = \sum_{i=1}^{N} a_i \phi_i \tag{2.43}$$

is the best approximation of ϕ obtainable with the trial functions ϕ_i. All trial function methods are concerned with this type of problem. The variational method of seeking an answer proceeds as follows.

Consider the one-dimensional case $\phi = \phi(x)$. The parameters a_i are constants. Assume that we know a functional V where

$$V = \int_{x_0}^{x_1} F(x, \phi)\, dx \tag{2.44}$$

whose Euler equation is Eq. (2.42). Then substitution of $\overline{\phi}$ for ϕ into Eq. (2.44) yields

$$\overline{V} = \int_{x_0}^{x_1} F(x, a_1, a_2, \ldots, a_N)\, dx \tag{2.45}$$

where $\overline{V}$ is an approximation of V.

Since V is stationary with respect to ϕ, we require $\overline{V}$ to be stationary with respect to $\overline{\phi}$ (i.e., with respect to $a_1, a_2, \ldots, a_N$). Accordingly, we require that

$$\delta\overline{V} = \frac{\partial \overline{V}}{\partial a_1} \delta a_1 + \frac{\partial \overline{V}}{\partial a_2} \delta a_2 + \cdots + \frac{\partial \overline{V}}{\partial a_N} \delta a_N = 0 \tag{2.46}$$

Since the a_i are arbitrary, Eq. (2.46) yields the N equations

$$\frac{\partial \overline{V}}{\partial a_i} = 0 \qquad i = 1, 2, \ldots, N \tag{2.47}$$

for the N parameters a_i. In this way we obtain an approximate solution to Eq. (2.42) that is the best in that it renders $\overline{V}$ stationary. In addition, we obtain a value of V that is accurate to second order in the error in the approximate solution for ϕ. Hence, if V_0 is the exact value of V, and if $\delta\phi$ is the error in the approximate solution of ϕ (i.e., $\overline{\phi} = \phi + \delta\phi$), then

$$\overline{V} = V_0 + O(\delta\phi^2) \tag{2.48}$$

The concept can be extended directly to multidimensional problems (Langhaar, 1989, Chap. 3). Often, in the multidimensional case, a semidirect approach is employed (Kantorovich and Krylov, 1964). For example, for two-dimensional problems $\phi = \phi(x, y)$, Eq. (2.43) is replaced by

$$\overline{\phi} = \sum a_i(y)\phi_i(x) \tag{2.49}$$

where now the parameters a_i are unknown functions of y. Then substitution of Eq. (2.49) into a functional

$$V = \int_{x0}^{x_1} dx \int_{y0}^{y_1} F(x, y, \phi) \, dy \tag{2.50}$$

yields

$$\overline{V} = \int_{y0}^{y_1} G(y, a_1, a_2, \ldots, a_N) \, dy \tag{2.51}$$

Then, setting the variation of $\overline{V}$ equal to zero, we obtain a set of ordinary differential equations that must be solved for the $a_i(y)$.

In statical elasticity problems, the functional V is the potential energy of the system, and the Euler equations are the equations of equilibrium. By the approximation method outlined above, a mechanical system (an elastic body, say) with infinitely many degrees of freedom is reduced to a system with finite degrees of freedom, since the a_i, $i = 1, 2, 3, \ldots, N$ may be thought of as generalized coordinates (Langhaar, 1989, Sec. 3.11). Rayleigh (1976) employed this idea in the study of vibrations of elastic bodies. Ritz (1909) refined and generalized Rayleigh's method, Ritz's method being equivalent to the procedure that leads to Eq. (2.47). In elasticity problems the condition $\delta V = 0$ is called the *principle of stationary potential energy.* Ritz's method approximates the condition $\delta V = 0$ by the conditions $\partial \overline{V} / \partial a_i = 0$, $i = 1, 2, \ldots, N$. These equations determine the a_i. Consequently, an approximation of ϕ is obtained by Eq. (2.43).

Example 2.6: Rayleigh–Ritz Method Applied to a Simple Beam The Rayleigh–Ritz method may be employed to approximate a continuous system by a system with a finite number of degrees of freedom. By way of illustration, consider a simply-supported beam of length L with lateral point load P applied at its midsection. The Bernoulli–Euler beam theory leads to the result

$$y(x) = \frac{PL^2 x}{16EI} - \frac{Px^3}{12EI} \tag{2.52}$$

for the deflection $y(x)$, where x denotes the axial coordinate from one end of the beam, E denotes the modulus of elasticity, and I is the moment of inertia of the cross-sectional area of the beam.

The internal potential energy (strain energy) of the beam due to bending is

$$U = \frac{1}{2} \int_0^L EI(y'')^2 \, dx \tag{2.53}$$

where the primes denote differentiation with respect to x. The potential energy of external loads is

$$\Omega = -Py\left(\frac{L}{2}\right) \tag{2.54}$$

Hence, the total potential energy of the beam is

$$V = U + \Omega = -Py\left(\frac{L}{2}\right) + \frac{1}{2} \int_0^L EI(y'')^2 \, dx \tag{2.55}$$

As a one-degree-of-freedom approximation, we let the deflection $y(x)$ be approximated by

$$\bar{y} = a \sin \frac{\pi x}{L} \tag{2.56}$$

where a is an unknown parameter. Substitution of Eq. (2.56) into Eq. (2.55) yields an approximation for V, namely,

$$\bar{V} = -aP + \frac{\pi^4 a^2 EI}{4L^3} \tag{2.57}$$

Substitution of Eq. (2.57) into (2.47) yields

$$a = \frac{2PL^3}{\pi^4 EI} \tag{2.58}$$

Hence,

$$\bar{y} = \frac{2PL^3}{\pi^4 EI} \sin \frac{\pi x}{L} \tag{2.59}$$

Equation (2.59) represents a fairly good approximation of $y(x)$, Eq. (2.52), as shown in Table 2.1. However, the derivatives of y are not approximated as accurately by derivatives of $\bar{y}$. For example, the bending moment of the beam is related to the second derivative of y by the relation

$$M = -EIy'' \tag{2.60}$$

or by Eq. (2.52),

$$M = \tfrac{1}{2}Px$$

Alternatively, substitution of $\bar{y}''$ for y'' in Eq. (2.60) yields the approximation $\bar{M}$ of M:

TABLE 2.1

x/L	EIy/PL^3	$EI\bar{y}/PL^3$	Error (%)
0	0	0	0
0.25	0.01432	0.01422	0.7
0.50	0.0208	0.0202	2.9

$$\overline{M} = \frac{2PL}{\pi^2} \sin \frac{\pi x}{L} \tag{2.61}$$

Hence, $M(L/2) = 0.25PL$, whereas $\overline{M}(L/2) = 0.203PL$, an error of approximately 19%. This result is fairly typical of approximations of the function y by $\overline{y}$ (primary functions), the higher derivatives of y (derived functions) being less accurately approximated by those of $\overline{y}$ than y is approximated by $\overline{y}$.

2.6   RITZ METHOD REVISITED AND TREFFTZ METHOD

As shown in Section 7.10 of Boresi and Chong (2000), the torsion problem for a simply connected region is represented by the equations

$$\begin{aligned} \nabla^2 \phi &= -2G\beta \qquad \text{over } R \\ \phi &= 0 \qquad \text{on } C \end{aligned} \tag{2.62}$$

In terms of the stress function ϕ, the stress components τ_{xz}, τ_{yz} are given by the equations

$$\tau_{xz} = \frac{\partial \phi}{\partial y} \qquad \tau_{yz} = -\frac{\partial \phi}{\partial x} \tag{2.63}$$

and the twisting moment M by

$$M = 2 \iint \phi \, dx \, dy \tag{2.64}$$

Equation (2.64) is essentially a boundary condition over the end planes (see Borcsi and Chong, 2000, Sec. 7.3).

Ritz Method Applied to a Rectangular Cross Section

We seek to obtain an approximate solution to the torsion problem of a rectangular cross section $-a \leqslant x \leqslant a$, $-b \leqslant y \leqslant b$ (Boresi and Chong, 2000, Sec. 7.10) by the Ritz method. This method (Section 2.5) is an interior method (Section 2.3) in that trial functions $\overline{\phi}$ are chosen to satisfy the boundary conditions. The arbitrary constants in the trial functions are then selected to provide a minimum for the integral over the cross section of the square of the error gradient (Ritz, 1909; Timoshenko and Goodier, 1970, Sec. 3). Thus, it may be shown that the error (residual) integral is minimized if

$$\iint_R \text{grad}^2 \, \overline{\phi} \, dx \, dy = \text{minimum} \qquad (2.65)$$

provided $\overline{\phi}$ is such that the boundary conditions [the second of Eqs. (2.62) and (2.64)] are satisfied.

The trial function $\overline{\phi}$ is chosen to satisfy the boundary condition on C (i.e., $\overline{\phi} = 0$ on C). The boundary condition on the end planes [Eq. (2.64)] may be introduced into Eq. (2.65) by means of the Lagrange multiplier method (see Boresi and Chong, 2000, Secs. 1.29 and 2.11). Thus, we define the auxiliary function to include the end plane boundary condition

$$F = \iint_R [\text{grad}^2 \, \overline{\phi} - L\overline{\phi}] \, dx \, dy \qquad (2.66)$$

where L is the Lagrange multiplier. The problem now is to determine $\overline{\phi}$ such that F is minimal and $\overline{\phi} = 0$ on C.

We may show that the problem of minimizing the function F is equivalent to minimizing the potential energy of the torsional system. We first note that the strain energy U in torsion is given by the equation [see Boresi and Chong, 2000, Eq. (4.6.13)]

$$U = \frac{1}{2G} \iint_R (\tau_{xz}^2 + \tau_{yz}^2) \, dx \, dy \qquad (2.67)$$

Substitution of Eqs. (2.63) into (2.67) yields (in terms of the trial function $\overline{\phi}$)

$$U = \frac{1}{2G} \iint_R \left[\left(\frac{\partial \overline{\phi}}{\partial y} \right)^2 + \left(\frac{\partial \overline{\phi}}{\partial x} \right)^2 \right] dx \, dy$$

$$= \frac{1}{2G} \iint_R \text{grad}^2 \, \overline{\phi} \, dx \, dy \qquad (2.68)$$

The potential energy Ω of the twisting moment (the external load) is given by

$$\Omega = -M\beta \tag{2.69}$$

By Eqs. (2.64) and (2.69), we have (in terms of the trial function $\overline{\phi}$)

$$\Omega = -2 \iint_R \overline{\phi}\beta \, dx \, dy \tag{2.70}$$

Hence, the total potential energy $V = U + \Omega$ of the system is given by (in terms of the trial function $\overline{\phi}$)

$$2GV = \iint_R [\text{grad}^2 \, \overline{\phi} - 4G\beta\overline{\phi}] \, dx \, dy \tag{2.71}$$

Comparison of Eqs. (2.66) and (2.71) shows that minimizing F is equivalent to minimizing $2GV$ with $L = 4G\beta$.

To satisfy the lateral boundary conditions $\phi = 0$ on C (Fig 2.1; see also Fig. 7.10.1 of Boresi and Chong, 2000), we may take

$$\overline{\phi} = (x^2 - a^2)(y^2 - b^2) \sum_{m=0}^{M} \sum_{n=0}^{N} A_{mn}x^m y^n \tag{2.72}$$

Since ϕ is even in (x, y), only even values of m and n need be considered.

First Approximation

To simplify the calculations, we consider a square cross section ($a = b$). Then, as a first approximation, we may take

$$\overline{\phi}_1 = A_{00}(x^2 - a^2)(y^2 - a^2) \tag{2.73}$$

Substitution of Eq. (2.73) into (2.66) yields, after integration,

$$F = a^6 \left(\frac{256}{45} a^2 A_{00}^2 - \frac{16}{9} L A_{00} \right) \tag{2.74}$$

and

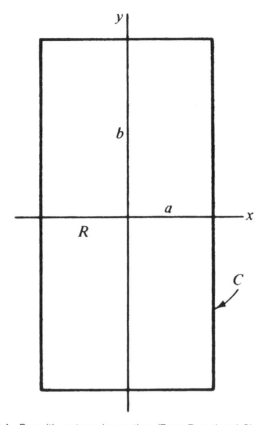

Figure 2.1   Bar with rectangular section. (From Boresi and Chong, 2000.)

$$\frac{\partial F}{\partial A_{00}} = \frac{512}{45} a^8 A_{00} - \frac{16}{9} a^6 L = 0 \tag{2.75}$$

With the condition $L = 4G\beta$, Eq. (2.75) yields the result

$$A_{00} = \frac{5G\beta}{8a^2} \tag{2.76}$$

and by Eqs. (2.63), (2.64), (2.73), and (2.76),

$$(\tau_{xz})_1 = \frac{5}{4} \frac{G\beta}{a^2} (x^2 - a^2)y$$

$$(\tau_{yz})_1 = -\frac{5}{4} \frac{G\beta}{a^2} (y^2 - a^2)x \tag{2.77}$$

$$M_1 = \frac{20}{9} G\beta a^4$$

Second Approximation

We next take [Eq. (2.72)]

$$\overline{\phi}_2 = [A_{00} + A_{22}(x^2 + y^2)](x^2 - a^2)(y^2 - a^2) \qquad (2.78)$$

Proceeding in the manner outlined above, we obtain the results

$$(\tau_{xz})_2 = \left[7770 + \frac{1575}{a^2}(x^2 - a^2 + 2y^2) \right] \frac{G\beta(x^2 - a^2)y}{6648a^2}$$

$$(\tau_{yz})_2 = -\left[7770 + \frac{1575}{a^2}(y^2 - a^2 + 2x^2) \right] \frac{G\beta(y^2 - a^2)x}{6648a^2} \qquad (2.79)$$

$$M_2 = 2.245G\beta a^2$$

Comparison of the results for the first and second approximations to the exact results (Boresi and Chong, 2000, Sec. 7.10) shows that the following hold for the Ritz method:

For given β:

$$M_{\text{exact}} > M_{N+1} > M_N > \cdots > M_2 > M_1 \qquad (2.80)$$

For given M:

$$\beta_{\text{exact}} < \beta_{N+1} < \beta_N < \cdots < \beta_2 < \beta_1 \qquad (2.81)$$

However, the comparisons of the stresses provide no such simple relations. For example, for $x = a$, $y = 0$,

$$\left|(\tau_{yz})_2\right| > \left|(\tau_{yz})_1\right|$$

whereas for $x = \sqrt{2}a/2$, $y = 0$,

$$\left|(\tau_{yz})_2\right| < \left|(\tau_{yz})_1\right|$$

The overall nature of the stress pattern is reflected by M and indicates that over a portion of the cross section, the stresses predicted are low in absolute value for a given value of twist β.

Trefftz Method Applied to a Rectangular Cross Section

In contrast to the Ritz method, the Trefftz method (Trefftz, 1927) is based on choosing trial functions $\overline{\phi}$ such that the differential equation [Eq. (2.62)] is satisfied over R, and the arbitrary constants in the trial functions are chosen to minimize the integral, over the cross section, of the error gradient. That is, the method is a *boundary method.* With this method the approximate magnitude of the twisting moment is larger than its exact value, in contrast to the Ritz method [Eq. (2.80)]. Hence, by using the Ritz method in conjunction with the Trefftz method, the exact value of the twisting moment M (hence the twist β) may be bounded from above and below.

Since the Trefftz method is a boundary method, we take the trial solution form

$$\overline{\phi} = -\tfrac{1}{2}G\beta(x^2 + y^2) + \sum_{i=1}^{N} B_i P_i \tag{2.82}$$

where $-G\beta(x^2 + y^2)/2$ is a particular solution of Eq. (2.62), the B_i are constants, and the P_i are potential functions ($\nabla^2 P_i = 0$).

If ϕ is the exact (true) solution, we may take the error gradient to be $\mathrm{grad}(\overline{\phi} - \phi)$. Thus, we require the minimization relative to B_i of

$$E = \iint_R \mathrm{grad}^2(\overline{\phi} - \phi)\, dx\, dy \tag{2.83}$$

that is

$$\frac{\partial E}{\partial B_i} = 2 \iint_R \mathrm{grad}(\overline{\phi} - \phi)\, \mathrm{grad}\, \frac{\partial \overline{\phi}}{\partial B_i}\, dx\, dy = 0 \tag{2.84}$$

where by Eq. (2.82),

$$\frac{\partial \overline{\phi}}{\partial B_i} = P_i \tag{2.85}$$

Thus,

$$\iint_R \mathrm{grad}(\overline{\phi} - \phi)\,\mathrm{grad}(P_i)\,dx\,dy = 0 \tag{2.86}$$

or, with reference to Eq. (1.16.6) of Boresi and Chong (2000),

$$\int_c (\overline{\phi} - \phi)\,\frac{\partial P_i}{\partial n}\,dS - \iint_R (\overline{\phi} - \phi)\nabla^2 P_i\,dx\,dy = 0 \tag{2.87}$$

where n is the outward-directed normal vector on C. Since $\nabla^2 P_i = 0$, Eq. (2.87) reduces to

$$\int_C \overline{\phi}\,\frac{\partial P_i}{\partial n}\,dS = \int_C \phi\,\frac{\partial P_i}{\partial n}\,dS \tag{2.88}$$

Hence, since in this case $\phi = 0$ on C, our problem is further reduced to requiring that

$$\int_C \overline{\phi}\,\frac{\partial P_i}{\partial n}\,dS = 0 \tag{2.89}$$

Again, for simplicity, we consider the square cross section $a = b$, and as a first approximation, we take

$$\overline{\phi} = \overline{\phi}_1 = -\frac{G\beta}{2}(x^2 + y^2) + B_1(x^4 - 6x^2y^2 + y^4) \tag{2.90}$$

Thus, $P_1 = x^4 - 6x^2y^2 + y^4 = P$.

For the square cross section Eq. (2.89) becomes, with Eq. (2.90),

$$\int_C \overline{\phi}\,\frac{\partial P}{\partial n}\,dS = 2\int_{-a}^{a}\overline{\phi}\,\frac{\partial P}{\partial x}\bigg|_{x=a}\,dy + 2\int_{-a}^{a}\overline{\phi}\,\frac{\partial P}{\partial y}\bigg|_{y=a}\,dx = 0 \tag{2.91}$$

Substitution of Eq. (2.90) into (2.91) yields, after integration and solution for B_1,

$$\overline{\phi}_1 = -\frac{G\beta}{2}\left[x^2 + y^2 + \frac{7}{36a^2}(x^4 - 6x^2y^2 + y^4)\right] \tag{2.92}$$

and

$$\tau_{xz} = \frac{\partial \overline{\phi}_1}{\partial y} = -\frac{G\beta}{2}\left[2y + \frac{28}{36a^2}(y^3 - 3x^2y)\right]$$

$$\tau_{yz} = -\frac{\partial \overline{\phi}_1}{\partial x} = \frac{G\beta}{2}\left[2x + \frac{28}{36a^2}(x^3 - 3y^2x)\right] \tag{2.93}$$

Since Eq. (2.92) does not satisfy identically the condition $\phi = 0$ on C, Eq. (2.64) gives an inaccurate result for M. Hence, we compute M by means of the expression for M following Eq. (7.3.5) of Boresi and Chong (2000); that is,

$$M = -\iint_R \left(x\frac{\partial \overline{\phi}}{\partial x} + y\frac{\partial \overline{\phi}}{\partial y}\right) dx\, dy \tag{2.94}$$

Substitution of Eq. (2.92) into (2.94) yields

$$M = 2.253G\beta a^4 \tag{2.95}$$

This procedure may be continued by considering additional terms in ϕ [Eq. (2.82)].

Bounds on Torsion Solution

The Ritz solution gives an upper bound for β and a lower bound for M, whereas the Trefftz method provides a lower bound for the trial function $\overline{\phi}$ (Trefftz, 1928). Since $\overline{\phi}$ is proportional to β, the Trefftz method therefore gives a lower bound for β and an upper bound for M, that is,

$$\beta_{\text{Trefftz}} < \beta_{\text{exact}} < \beta_{\text{Ritz}}$$

$$M_{\text{Trefftz}} > M_{\text{exact}} > M_{\text{Ritz}}$$

In the example discussed above, the results are

$$0.4444\frac{M}{Ga^4} < \beta_{\text{exact}} < 0.445\frac{M}{Ga^4}$$

$$2.253G\beta a^4 > M_{\text{exact}} > 2.245G\beta a^4$$

giving very close bounds on the exact results.

For stresses the comparison is more difficult. However, comparisons for maximum stress can be given. For example, for the square cross section, τ_{max} occurs at the midpoint of a side, say, at $x = a$, $y = 0$.

Then, for the Ritz solution, Eqs. (2.77) and (2.79) yield $(\tau_{max})_1 = 1.25G\beta a$, $(\tau_{max})_2 = 1.41G\beta a$ and Eq. (2.93) yields for the Trefftz method $\tau_{max} = 1.39G\beta a$. The series solution (Boresi and Chong 2000, Sec. 7.10; Timoshenko and Goodier, 1970, p. 312) yields $\tau_{max} = 1.35G\beta a = (\tau_{max})_{exact}$. Unfortunately, the stresses are not bounded.

REFERENCES

Allen, M. B., III, and Isaacson, E. L. 1997. *Numerical Analysis for Applied Science,* Wiley, New York.

Bathe, K. J. 1995. *Finite Element Procedures,* Prentice Hall, Upper Saddle River, NJ.

Boresi, A. P., and Chong, K. P. 2000. *Elasticity in Engineering Mechanics,* Wiley, New York.

Botha, D. K., and Pinder, G. F. 1983. *Fundamental Concepts in the Numerical Solution of Differential Equations,* Wiley, New York.

Collatz, L. 1960. *The Numerical Treatment of Differential Equations,* 3rd ed., Springer-Verlag, Berlin.

Cook, R. D., Malkus, D. S., and Plesha, M. E. 1989. *Concepts and Applications of Finite Element Analysis,* 3rd ed., Wiley, New York.

Crandall, S. H. 1956. *Engineering Analysis,* Academic Press, San Diego, CA, p. 231.

Finlayson, B. A. 1972. *The Method of Weighted Residuals and Variational Principles,* Academic Press, San Diego, CA.

Finlayson, B. A., and Scriven, L. E. 1966. The method of weighted residuals: a review, *Appl. Mech. Rev.,* **19**(9).

Hughes, T. J. R. 2000. *The Finite Element Method: Linear Static and Dynamic Analysis,* Dover Publications, Mineola, NY.

Kantorovich, L., and Krylov, V. 1964. *Approximate Methods of Higher Analysis,* Wiley, New York.

Langhaar, H. L. 1989. *Energy Methods in Applied Mechanics,* Kreiger Publishing, Melbourne, FL.

Rayleigh, Lord (J. S.) 1976. *Theory of Sound,* Dover Publications, Mineola, NY.

Reddy, J. N. 1991. *Applied Functional Analysis and Variational Methods in Engineering,* Krieger Publishing, Melbourne, FL.

Ritz, W. 1909. Uber eine neue Methode zur Lösung gewisser Variations probleme der mathematischen Physik, *J. Reine Angew. Math.,* **135,** 1–61.

Shisha, O. 1968. Trends in approximation theory, *Appl. Mech. Rev.,* **21,** 4.

Timoshenko, S. P., and Goodier, J. N. 1970. *Theory of Elasticity,* 3rd ed., McGraw-Hill, New York.

Trefftz, E. 1927. Ein Gegenstück zum Ritzschen Verfahren, *Verh. Zweiten Kongress Tech. Mech.* (Zurich), pp. 131–137.

Trefftz, E. 1928. Konvergenz und Fehlerschätzung beim Ritzschen Verfahren, *Math. Ann.,* **100,** 503–521.

Yang, H. Y., and Zhao, Z. G. 1985. The mixed method of line-location with its applications in elasticity, *Comput. Struct.,* **21,** 671–680.

Zienkiewicz, O. C., and Morgan, K. 1983. *Finite Elements and Approximation,* Wiley, New York.

BIBLIOGRAPHY

Brebbia, C. A. (ed.) 1983. *Progress in Boundary Element Methods,* Vol. 2 Springer-Verlag, New York.

Cakmak, A. S., and Botha, J. F. 1995. *Applied Mathematics for Engineers,* Computational Mechanics, Billerica, MA.

Dhatt, G., and Touzot, G. 1984. *The Finite Element Method Displayed,* Wiley, New York.

Finlayson, B. A. 1992. *Numerical Methods for Problems with Moving Fronts,* Ravenna Park Publishing, Seattle, WA.

Langhaar, H. L. 1969. Two numerical methods that converge to the method of least squares, *J. Franklin Inst.,* **288,** 165–173.

Lapidus, L., and Pinder, G. F. 1982. *Numerical Solution of Partial Differential Equations in Science and Engineering,* Wiley, New York.

Mikhlin, S. G. 1964. *Variational Methods in Mathematical Physics,* Macmillan, New York.

Mikhlin, S. G., and Smolitskii, K. L. 1967. *Approximate Methods for Solution of Differential and Integral Equations,* Elsevier Science, New York.

Morse, P. M., and Feshbach, H. 1953. *Methods of Theoretical Physics,* McGraw-Hill, New York, pp. 122–123.

Yang, T. Y. 1986. *Finite Element Structural Analysis,* Prentice Hall, Upper Saddle River, NJ.

3

Finite Difference Methods

3.1 PRELIMINARY REMARKS AND CONCEPTS

The basic concept in finite difference methods, as applied to boundary-value problems, is the representation of governing differential equations and associated boundary conditions by appropriate finite difference equations. This replacement is accomplished by approximating derivatives in the differential equations with finite difference quotients that are combinations of dependent (unknown) function values at specified values of the independent variables. By writing the difference equations at specified values of the independent variables, we are led to systems of simultaneous algebraic equations that may be solved by elementary means with the aid of high-speed computers. Accordingly, we interpret a finite difference method as a numerical procedure that approximates known *exact*[1] differential equations and boundary conditions—say, of an elasticity problem. Then we solve the resulting approximate equations exactly or approximately. On the other hand, we shall see in Chapter 4 that finite element methods approximate the elastic continua by assemblages of discrete elastic systems. Then we solve the resulting discrete systems exactly or approximately.

To illustrate the approximation of derivatives by corresponding difference quotients, consider the one-dimensional case. Let $f(x)$ be a

---

[1] The term *exact* is used in a sense that the differential equations are derived exactly under the framework of the basic assumptions of the theory.

continuous function of the single independent variable x, which is defined in the range $a_0 \leqslant x \leqslant a_n$ or $[a_0, a_n]$. We divide $[a_0, a_n]$ into n equal or unequal parts and denote the subdividing points as $a_0, a_1, a_2,$ $\ldots, a_n$. These points are called *pivotal points*. From the theory of interpolation polynomials, we construct an interpolation polynomial P_n of degree n satisfying the conditions $f(a_i) = P_n(a_i)$, $i = 0, 1, 2, \ldots,$ n. Such an interpolation polynomial is expressible as a linear combination of function values $f(a_1)$ (Isaacson and Keller, 1966). Accordingly, we write $f(x) = P_n(x) + R_n(x)$ for all x in $[a_0, a_n]$, where $R_n(x)$ is a remainder term or simply the error in representation of $f(x)$ by $P_n(x)$. Since the derivatives $d^r f(x)/dx^r = d^r P_n(x)/dx^r + d^r R_n(x)/dx^r$, we use the derivatives of the interpolation polynomial to replace approximately the derivatives of the interpolated function $f(x)$.

If the derivatives of the error terms can be estimated, they provide a means of judging the accuracy of the approximation. Upon suppressing the error term, we have the approximation $d^r f(x)/dx^r \approx d^r P_n(x)/dx^r$, where now the $P_n(x)$ is no longer expressed in terms of exact pivotal function values $f(a_i)$, but rather in terms of a linear combination of approximation values of $f(a_i)$, which we denote by $F(a_i)$. Proceeding in this manner, we express all derivatives in the governing differential equations and the boundary conditions in terms of finite difference equations. Evaluation of these finite difference equations at the specified pivotal points in the domain of a boundary-value problem leads to a set of algebraic equations in $F(a_i)$. For linear boundary-value problems, these algebraic equations form a system of linear simultaneous equations. Otherwise, they form a system of nonlinear algebraic equations.

Finite Differences, Finite Elements, and Weighted Residual Methods

Much effort has been directed toward the unification of various approximation processes used in the numerical solution of physical problems governed by suitable differential equations (Bushnell, 1973; Zienkiewicz and Morgan, 1983). The finite difference process discussed in this chapter appears to present an entirely different type of approach from that of the weighted residual methods of Chapter 2 and also from the finite element methods of Chapter 4. However, Zienkiewicz and Morgan (1983) have shown that in several one- and two-dimensional cases, the use of finite differences has led to equations that are either identical or extremely similar to those obtained by simple finite elements. They have also shown that the common link between these approximation processes is expansion of the unknown function in terms of shape or basis functions and unknown parameters and the determination of such parameters from a set of weighted residual equations. They coined the phrase *generalized finite element method* to include all the approaches mentioned. With such a generalized approach, computer program organization and the theory to encompass all the approximation processes are unified. Furthermore, arguments concerning the superiority of finite difference methods over finite element methods (or vice versa) become meaningless, as each subclass possesses its own merits in special circumstances.

3.2 DIVIDED DIFFERENCES AND INTERPOLATION FORMULAS

In the theory of interpolating polynomials, *divided differences* play an important role. In this section we present a few useful properties of divided differences. Consider arbitrary pivotal points $a_0, a_1, a_2, \ldots, a_n$ of the independent variable x and the corresponding function values $f(a_0), f(a_1), f(a_2), \ldots, f(a_n)$. The zero-order divided difference is defined as $f[a_0] = f(a_0)$. First-, second-, and higher-order divided differences are defined by

$$f[a_0, a_1] = \frac{f(a_1) - f(a_0)}{a_1 - a_0} = \frac{f(a_0)}{a_0 - a_1} + \frac{f(a_1)}{a_1 - a_0}$$

$$f[a_0, a_1, a_2] = \frac{f[a_1, a_2] - f[a_0, a_1]}{a_2 - a_0} = \frac{f(a_0)}{(a_0 - a_1)(a_0 - a_2)}$$

$$+ \frac{f(a_1)}{(a_1 - a_0)(a_1 - a_2)} + \frac{f(a_2)}{(a_2 - a_0)(a_2 - a_1)}$$

$$\vdots$$

$$f[a_0, a_1, a_2, \ldots, a_n] = \frac{f[a_1, a_2, \ldots, a_n] - f[a_0, a_1, a_2, \ldots, a_{n-1}]}{a_n - a_0}$$

$$= \sum_{j=0}^{n} \frac{f(a_j)}{(a_j - a_0)(a_j - a_1) \cdots (a_j - a_{j-1})(a_j - a_{j+1}) \cdots (a_j - a_n)}$$

$$(3.1)$$

By Eqs. (3.1), note that the divided differences remain invariant with respect to any permutation of pivotal points and are simply linear combinations of the pivotal function values $f(a_j)$ at the pivotal points a_0, $a_1, \ldots, a_n$.

The basic form of an interpolation formula due to Newton may be derived directly from the definitions of divided differences. Consider the first-order divided difference with arguments a_0 and x:

$$f[a_0, x] = \frac{f(x) - f(a_0)}{x - a_0}$$

Hence,

$$f(x) = f[a_0] + (x - a_0)f[a_0, x] \tag{3.2}$$

By expanding expressions $f[a_1, a_0, x]$, $f[a_2, a_0, a_1, x]$, . . . and $f[a_n, a_0, \ldots, a_{n-1}, x]$ according to Eqs. (3.1), we further have

$$f[a_0, x] = f[a_0, a_1] + (x - a_1)f[a_0, a_1, x]$$

$$\vdots \tag{3.3}$$

$$f[a_0, a_1, \ldots, a_{n-1}, x] = f[a_0, a_1, \ldots, a_n]$$
$$+ (x - a_n)f[a_0, a_1, \ldots, a_n, x]$$

Successive substitution of Eqs. (3.3) into (3.2) yields

$$f(x) = f[a_0] + (x - a_0)f[a_0, a_1] + (x - a_0)(x - a_1)f[a_0, a_1, a_2]$$
$$+ \cdots + (x - a_0)(x - a_1)(x - a_2) \cdots$$
$$(x - a_{n-1})f[a_0, a_1, \ldots, a_n] + R_n(x) \tag{3.4}$$

where $R_n(x)$, the remainder or error term, is given by

$$R_n(x) = \omega_n(x)f[a_0, a_1, \ldots, a_n, x] \tag{3.5}$$

where

$$\omega_n(x) = (x - a_0)(x - a_1) \cdots (x - a_n) \tag{3.6}$$

Equation (3.4) may be written in the form

$$f(x) = P_n(x) + R_n(x) \qquad (3.7)$$

where $P_n(x)$ is the nth degree (Newton) interpolation polynomial for the function $f(x)$ in terms of divided differences. Accordingly, $P_n(x)$ depends on the $(n + 1)$ pivotal function values at the $(n + 1)$ distinct points, 0, 1, 2, . . . , n. Suppressing the error term in Eq. (3.7), we obtain the approximation $f(x) \approx P_n(x)$. If an approximate representation of the rth $(r \leqslant n)$ derivative of $f(x)$ is desired, we may differentiate Eq. (3.7) r times to obtain

$$\frac{d^r f(x)}{dx^r} \approx \frac{d^r P_n(x)}{dx^r} \quad \text{or simply} \quad f^{(r)}(x) \approx P_n^{(r)}(x) \qquad (3.8)$$

with the associated error $R_n^{(r)}(x)$, where the r superscript in parentheses denotes the order of differentiation. An estimate of the error term $R_n^{(r)}(x)$ may reveal the accuracy and the rate of convergence of the approximation.

For assessment of the error, a convenient representation of $R_n^{(r)}(x)$ is required. For this representation we note two important properties of divided differences. They are stated as follows (with brief proofs).

Property I   Let $a_0, a_1, \ldots, a_{p-1}, x$ be $(p + 1)$ distinct pivotal points, and let $f(x)$ possess p continuous derivatives in the closed interval $I = [a_0, x]$ contained between the smallest and largest pivotal points. Then there exists a point $\xi = \xi(x)$ in the interval I such that

$$f[a_0, a_1, \ldots, a_{p-1}, x] = \frac{f^{(p)}(\xi)}{(p)!} \qquad (3.9)$$

Proof   Denote by $P_{p-1}(x)$ the $(p - 1)$th degree interpolation polynomial for $f(x)$ with respect to the p pivotal points $a_0, a_1, \ldots, a_{p-1}$. Then $f(x) - P_{p-1}(x)$ and $\omega_{p-1}(x)$ (see Eq. (3.6)) vanish at the p points. Define a linear combination of these functions as

$$G(x) = f(x) - P_{p-1}(x) - \alpha \omega_{p-1}(x) \qquad (3.10)$$

and note that $G(x_i) = 0$ for $x_i = a_i$ $(i = 0, 1, 2, \ldots, p - 1)$. For any arbitrary value x distinct from a_i, $\omega_{p-1}(x) \neq 0$; hence, we may select the constant α so that $G(x)$ vanishes. Since $G(x_i) = 0$ at $x_i = a_0, a_1,$ . . . , a_{p-1}, x, there are $(p + 1)$ zeros for $G(x)$ in the interval I. By Rolle's theorem (Thomas, 1960), $dG(x)/dx$ vanishes at least p times inside I. Repeated application of Rolle's theorem shows that $d^2 G(x)/$

$dx^2 = 0$ at least $(p - 1)$ times and $d^r G(x)/dx^r = 0$ at least $(p - r + 1)$ times inside $I(r \leq p)$. Accordingly, there must be a point $\xi = \xi(x)$ inside the interval I that yields $G^p(\xi) = 0$. Equation (3.10) then yields

$$\alpha = \frac{f^{(p)}(\xi) - P^{(p)}_{p-1}(\xi)}{\omega^{(p)}_{p-1}(\xi)}$$

The fact that $P_{p-1}(x)$ is a polynomial of degree $(p - 1)$ at most and the fact that $\omega_{p-1}(x)$ is a polynomial of degree p show, respectively, that $P^{(p)}_{p-1}(\xi) = 0$ and $\omega^{(p)}_{p-1}(\xi) = p!$. Thus the constant α becomes $\alpha = f^{(p)}(\xi)/p!$. Since we have required $G(x) = 0$ at the point x in I, it follows from Eqs. (3.7) and (3.10) that

$$\omega_{p-1}(x) \frac{f^{(p)}(\xi)}{p!} = f(x) - P_{p-1}(x) \tag{3.11}$$

On the basis of Eqs. (3.5) and (3.11), (3.9) is proved. Furthermore, note that Eq. (3.11) holds for previously excluded values of x corresponding to $a_0, a_1, \ldots, a_{p-1}$. Consequently, Eq. (3.9) is valid for all x.

Property II Let $f^{(p)}(x)$ be continuous in the closed interval I, and let $a_0, a_1, \ldots, a_n, x$ be in I. Then

$$\frac{d^p}{dx^p} f[a_0, a_1, a_2, \ldots, a_n, x] = (p!)f[a_0, a_1, a_2, \ldots, a_n, x, x, \ldots, x] \tag{3.12}$$

where x is repeated $p + 1$ times on the right-hand side.

Proof Consider a function

$$g(x) = f[a_0, a_1, a_2, \ldots, a_n, x] \tag{3.13}$$

By definition $g^{(p)}(x)$ is also continuous in I. Letting $b_0, b_1, b_2, \ldots, b_p$ be another set of $p + 1$ points in I, we form (Hartree, 1961)

$$g[b_0, b_1] = \frac{g(b_1) - g(b_0)}{b_1 - b_0}$$

$$= \frac{f[a_0, a_1, \ldots, a_n, b_1] - f[a_0, a_1, \ldots, a_n, b_0]}{b_1 - b_0}$$

$$= f[a_0, a_1, a_2, \ldots, a_n, b_0, b_1]$$

$$\vdots$$

$$g[b_0, b_1, \ldots, b_p] = f[a_0, a_1, \ldots, a_n, b_0, b_1, \ldots, b_p] \qquad (3.14)$$

On the basis of Eq. (3.9), it follows, by taking b_p as x, that

$$\frac{g^{(p)}(\xi)}{p!} = f[a_0, a_1, a_2, \ldots, a_n, b_0, b_1, \ldots, b_p] \qquad (3.15)$$

where ξ is inside the interval I. On account of the continuity of $g^{(p)}(x)$ in the interval I, we let $x = b_0 = b_1 = \cdots = b_p$. Thus Eq. (3.15) reduces to (see Appendix 3A)

$$\frac{g^{(p)}(x)}{p!} = f[a_0, a_1, a_2, \ldots, a_n, x, x, \ldots, x] \qquad (3.16)$$

where x is repeated $p + 1$ times on the right-hand side. Finally, by means of Eqs. (3.13) and (3.16), we deduce Eq. (3.12).

Returning to the error term associated with the approximation of Eq. (3.8), we note that

$$R_n^{(r)}(x) = \frac{d^r}{dx^r} \omega_n(x) f[a_0, a_1, a_2, \ldots, a_n, x] \qquad (3.17)$$

Differentiating the product of two functions according to Leibnitz's rule, we obtain

$$R_n^{(r)}(x) = \sum_{i=0}^{r} \binom{r}{i} \frac{d^i}{dx^i} \omega_n(x) \frac{d^{r-i}}{dx^{r-i}} f[a_0, a_1, a_2, \ldots, a_n, x]$$

where

$$\binom{r}{i} = \frac{r!}{(r-i)!i!}$$

In view of Eq. (3.12), we write

$$R_n^{(r)}(x) = \sum_{i=0}^{r} \frac{r!}{i!} \, \omega_n^{(i)}(x) f[a_0, a_1, a_2, \dots, a_n, x, x, \dots, x] \quad (3.18)$$

where x is repeated $(r - i + 1)$ times on the right-hand side. Now a generalization of Eq. (3.9) yields

$$f[a_0, a_1, a_2, \dots, a_n, x, x, \dots, x] = \frac{f^{(n+r-i+1)}(\xi_i)}{(n+r-i+1)!} \quad (3.19)$$

where $\min(a_0, a_1, \dots, a_n, x) < \xi_i(x) < \max(a_0, a_1, \dots, a_n, x)$, and where x is repeated $r - i + 1$ times on the left-hand side.

Accordingly, substitution of Eq. (3.19) into Eq. (3.18) leads to

$$R_n^{(r)}(x) = \sum_{i=0}^{r} \frac{r!}{(n+r-i+1)!i!} \, \omega_n^{(i)}(x) f^{(n+r-i+1)}(\xi_i) \quad (3.20)$$

Since the behavior of higher derivatives of $f(x)$ is generally unknown, some assumptions regarding boundedness of these derivatives are necessary in estimating the error term. The factor $\omega_n^{(i)}(x)$ is associated with the spacings of the pivotal points. Hence, it serves as an indicator of vanishing error with respect to diminishing pivotal spacings.

Expressions for Newton's interpolation polynomial and its associated error term [Eq. (3.4)] are derived in terms of divided differences for the interpolation interval $[a_0, a_n]$. They are equally applicable to both nonuniform and uniform spacings of the pivotal points. *However, in practice, uniform spacing is widely used.* Accordingly, we present some useful interpolation polynomials for equally spaced pivotal points, where the constant interval (spacing) is denoted by h.

Generally, there are three types of interpolation polynomials, depending on where the initial interpolating point a_0 is placed in the interval of interpolation $[a_0, a_n]$. For example, if we evaluate $P_n(x)$ near the right-hand region of the interpolation interval, it is natural to place a_0 close to the right-hand endpoint, so that greater emphasis may be placed on the pivotal function values near this end. Consequently, we introduce three sets of differently ordered pivotal points for *forward, backward,* and *central* interpolation polynomials, as shown in Fig. 3.1a, b, and c, respectively. Figure 3.1 also indicates the definition of new systems of pivotal coordinates $x_i = x_0 + ih$ ($i = 0, \pm 1, \pm 2, \ldots$). This notation enables us to write any pivotal function value $f(x_i)$ in a convenient subscripted form [i.e., $f(x_i) = f(x_0 + ih) = f_{0+i}$].

Corresponding to the forward, backward, and central arrangements of the pivotal points shown in Fig. 3.1, we introduce *forward, backward,* and *central differences* as follows:

$$\Delta f(x_i) = f(x_1 + h) - f(x_i) = f_{i+1} - f_i \qquad \text{(forward)}$$

$$\nabla f(x_i) = f(x_i) - f(x_i - h) = f_i - f_{i-1} \qquad \text{(backward)}$$

$$\delta f(x_i) = f(x_i + \tfrac{1}{2}h) - f(x_i - \tfrac{1}{2}h) = f_{i+1/2} - f_{i-1/2} \quad \text{(central)}$$

$$(3.21)$$

Note that the first central difference entails values of $f(x)$ other than those corresponding to values of x_i. In a manner similar to first-order differences, we write formulas for higher-order differences. For example, second-order differences are defined by

$$\Delta^2 f(x_i) = \Delta(\Delta f(x_i)) = \Delta(f_{i+1} - f_i) = f_{i+2} - 2f_{i+1} + f_i \qquad \text{(forward)}$$

$$\nabla^2 f(x_i) = \nabla(\nabla f(x_i)) = \nabla(f_i + f_{i-1}) = f_i - 2f_{i-1} + f_{i-2} \qquad \text{(backward)}$$

$$\delta^2 f(x_i) = \delta(\delta f(x_i)) = \delta(f_{i+1/2} - f_{i-1/2}) = f_{i+1} - 2f_i + f_{i-1} \quad \text{(central)}$$

$$(3.22)$$

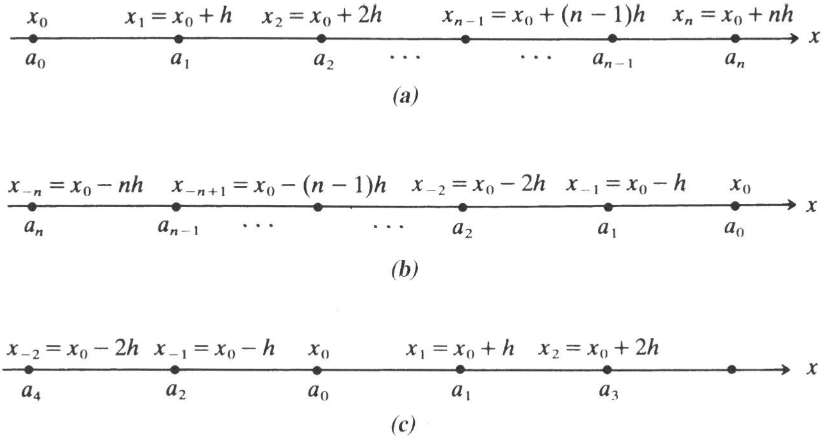

Figure 3.1 Equal pivotal spacings.

We see that the second-order central difference depends on values of f_i. In general, central even differences depend on values of f_i, and central odd differences depend on $f_{i+1/2}$. Generalization of Eqs. (3.21) and (3.22) yields for $r = 1, 2, 3, \ldots$,

$$\Delta^r f_i = \Delta(\Delta^{r-1} f_i) \qquad \text{(forward)}$$

$$\nabla^r f_i = \nabla(\nabla^{r-1} f_i) \qquad \text{(backward)} \qquad (3.23)$$

$$\delta^r f_i = \delta(\delta^{r-1} f_i) \qquad \text{(central)}$$

with $\Delta^0 f_i = \nabla^0 f_i = \delta^0 f_i = f_i$.

Newton's Forward and Backward Interpolation Polynomials

Consider the case of equally spaced x_i (Fig. 3.1a). Newton's forward interpolation polynomial may be derived with respect to point x_0 using pivotal points to the right of point x_0. We first write the divided differences of Eq. (3.1) directly in terms of the foward differences:

$$f[x_0, x_1] = \frac{1}{h}(f_1 - f_0) = \frac{1}{h}\Delta f_0$$

$$f[x_0, x_1, x_2] = \frac{1}{2h}(f[x_1, x_2] - f[x_0, x_1]) = \frac{1}{2!h^2}\Delta^2 f_0$$

$$\vdots$$

$$f[x_0, x_1, x_2, \ldots, x_n] = \frac{1}{nh}(f[x_1, x_2, \ldots, x_n] - f[x_0, x_1, \ldots, x_{n-1}])$$

$$= \frac{1}{n!h^n}\Delta^n f_0 \qquad (3.24)$$

On denoting x by $x_0 + sh$, where s measures $x - x_0$ in units of h, and eliminating the divided differences by means of Eqs. (3.24) from the general term of Newton's interpolation polynomial defined in Eq. (3.4), we obtain

$$(x - x_0)(x - x_1) \cdots (x - x_{n-1}) f[x_0, x_1, \ldots, x_n]$$

$$= s(s - 1)(s - 2) \cdots (s - n + 1) \frac{\Delta^n f_0}{n!} \tag{3.25}$$

Hence, Newton's forward interpolation polynomial $P_n(x) = P_n(x_0 + sh)$ is expressible as

$$P_n(x_0 + sh) = f_0 + s \, \Delta f_0 + s(s - 1) \frac{\Delta^2 f_0}{2!} + \cdots$$

$$+ s(s - 1) \cdots (s - n + 1) \frac{\Delta^n f_0}{n!} \tag{3.26}$$

The rth $(r \leq n)$ derivative of $P_n(x_0 + sh)$ with respect to s then yields the desired approximation to $f^{(r)}(x_0 + sh)$. The associated error can be easily estimated from Eq. (3.20), as is demonstrated later in connection with finite difference equations.

If the order of x_i is reversed so that $n - 1$ interpolation points are to the left of x_0 as shown in Fig. 3.1b, we may construct Newton's backward interpolation polynomial $P_n(x_0 - sh)$. In a manner analogous to the previous derivation, we obtain

$$P_n(x_0 - sh) = f_0 + (-s) \nabla f_0 + (-s)(-s + 1) \frac{\nabla^2 f_0}{2!} + \cdots$$

$$+ (-s)(-s + 1) \cdots (-s + n - 1) \frac{\nabla^n f_0}{n!} \tag{3.27}$$

The forward and backward differences in Eqs. (3.26) and (3.27) can be conveniently expressed in terms of linear combinations of pivotal function values f_i. For example, note from Eq. (3.1) that

$$f[x_0, x_1, x_2, \ldots, x_n] = \sum_{i=0}^{n} f_i \Big/ \prod_{\substack{j=0 \\ j \neq 1}}^{n} (x_i - x_j) \tag{3.28}$$

However,

$$\prod_{\substack{j=0 \\ j \neq i}}^{n} (x_i - x_j) = \prod_{\substack{j=0 \\ j \neq i}}^{n} (i - j)h - h^n \prod_{j=0}^{i-1} (i - j) \prod_{r=i+1}^{n} (i - r)$$

$$= h^n (i)! (n - i)! (-1)^{n-i}$$

Hence, Eq. (3.28) may be written in the form

$$f[x_0, x_1, x_2, \ldots, x_n] = \frac{1}{n! h^n} \sum_{i=0}^{n} (-1)^{n-i} \frac{n!}{(n-i)! \, i!} f_i$$

$$= \frac{1}{n! h^n} \sum_{i=0}^{n} (-1)^{n-i} \binom{n}{i} f_i \qquad (3.29)$$

where

$$\binom{n}{i} = \frac{n!}{(n-i)! \, i!}$$

On comparing the last equation of Eq. (3.24) with (3.29), we deduce that

$$\Delta^n f_0 = \sum_{i=0}^{n} (-1)^{n-i} \binom{n}{i} f_i \qquad (3.30)$$

In a similar manner, we find that

$$\nabla^n f_0 = \sum_{i=0}^{n} (-1)^i \binom{n}{i} f_i \qquad (3.31)$$

These formulas may be used to facilitate conversion of higher-order differences to their corresponding combinations of function values f_i.

Newton–Gauss Interpolation Polynomial

Many occasions arise in the application of finite difference techniques where it is required to approximate derivatives of $f(x)$ at interior pivotal points (Stanton, 1961; Chapra, 1990). To cope with these situations, we develop central interpolation polynomials. Let the arrangement of the pivotal points x_i be centrally distributed about the point x_0 as shown in Fig. 3.1c. We derive first the *Newton–Gauss interpolation formula*. From the definition of the divided differences of Eq. (3.1) and the centrally arranged pivotal points x_i, we write

$$f[x_0, x_1] = \frac{1}{h}(f_1 - f_0) = \frac{1}{h}\delta f_{0+1/2}$$

$$f[x_{-1}, x_0, x_1]$$

$$= \frac{f[x_0, x_1] - f[x_{-1}, x_0]}{x_1 - x_{-1}} = \frac{1}{2h^2}(\delta f_{0+1/2} - \delta f_{0-1/2}) = \frac{\delta^2 f_0}{2!h^2}$$

$$\vdots$$

$$f[x_{-m+1}, \ldots, x_0, \ldots, x_m] = \frac{\delta^{2m-1} f_{0+1/2}}{(2m-1)!h^{2m-1}}$$

$$f[x_{-m}, \ldots, x_0, \ldots, x_m] = \frac{\delta^{2m} f_0}{(2m)!h^{2m}} \qquad (3.32)$$

Next, consider the factors associated with the last two divided differences of Eqs. (3.32) as they appear to Newton's interpolation polynomial [Eq. (3.4)]. Noting that $x_i = x_0 + ih$ with $i = 0, \pm 1, \pm 2, \ldots, \pm m$, we write ($n = 2m - 1$)

$$\prod_{i=-m}^{m-1} (x - x_i) = (x - x_0) \prod_{i=-1}^{-m+1} (x - x_i) \prod_{i=1}^{m-1} (x - x_i)$$

$$= sh^{2m-1} \prod_{i=1}^{m-1} (s^2 - i^2) \qquad (3.33)$$

where $x = x_0 + sh$ is used. Similarly, we obtain

$$\prod_{i=-m+1}^{m} (x - x_i) = sh^{2m}(s - m) \prod_{i=1}^{m-1} (s^2 - i^2) \qquad (3.34)$$

Using Eqs. (3.32), (3.33), and (3.34), we rewrite Eq. (3.4) into the following form, known as the *Newton–Gauss central interpolation formula:*

$P_{2m}(x_0 + sh)$

$$= f_0 + s\delta f_{0+1/2} + s(s - 1)\frac{\delta^2 f_0}{2!} + s(s^2 - 1^2)\frac{\delta^3 f_{0+1/2}}{3!}$$

$$+ s(s^2 - 1^2)(s - 2)\frac{\delta^4 f_0}{4!} + \cdots$$

$$+ s\left[\prod_{i=1}^{m-1}(s^2 - i^2)\right]\frac{\delta^{2m-1} f_{0+1/2}}{(2m - 1)!}$$

$$+ s(s - m)\left[\prod_{i=1}^{m-1}(s^2 - i^2)\right]\frac{\delta^{2m} f_0}{(2m)!} \tag{3.35}$$

If the last term is omitted, we obtain the expression for $P_{2m-1}(x_0 + sh)$.

Stirling's Central Interpretation Polynomial

Because of the unsymmetrical nature of the odd-order central differences of Eq. (3.35), its practical use is limited. However, we may rearrange the terms and effect a symmetrical representation, called *Stirling's central interpolation formula,* that is commonly used in finite difference methods. In practice, it is natural to select an equal number of pivotal points to the right and to the left of the point x_0. Then we limit ourselves to even-degree interpolation polynomials $P_{2m}(x_0 + sh)$. We then rewrite Eq. (3.35) as

$P_{2m}(x_0 + sh)$

$$= f_0 + \left[s\delta f_{0+1/2} - \frac{s}{2!}\delta^2 f_0\right] + \frac{s}{2 \times 2!}[(s - 1) + (s + 1)]\delta^2 f_0$$

$$+ \left[s(s^2 - 1^2)\frac{\delta^2 f_{0+1/2}}{3!} - \frac{2s}{4!}\delta^4 f_0\right]$$

$$+ \frac{s(s^2 - 1^2)}{2 \times 4!}[(s - 2) + (s + 2)]\delta^4 f_0$$

$$+ \cdots$$

$$+ \left\{s\left[\prod_{i=1}^{m-1}(s^2 - i^2)\right]\frac{\delta^{2m-1} f_{0+1/2}}{(2m - 1)!} - \frac{sm}{(2m)!}\delta^{2m} f_0\right\}$$

$$+ \left[\prod_{i=1}^{m-1}(s^2 - i^2)\right]\frac{s[(s - m) + (s + m)]}{2 \times (2m)!}\delta^{2m} f_0$$

By expressing all the even-order central differences in the brackets as combinations of their next-lower central odd differences and then combining the terms within each bracket, we obtain Stirling's formula:

$$P_{2m}(x_0 + sh)$$

$$= f_0 + \frac{s}{2}(\delta f_{0+1/2} + \delta f_{0-1/2}) + \frac{s}{2 \times 2!}[(s-1) + (s+1)]\,\delta^2 f_0$$

$$+ \frac{s(s^2 - 1^2)}{2 \times 3!}(\delta^3 f_{0+1/2} + \delta^3 f_{0-1/2})$$

$$+ \frac{s(s^2 - 1^2)}{2 \times 4!}[(s-2) + (s+2)]\,\delta^4 f_0$$

$$+ \cdots$$

$$+ \frac{s}{2(2m-1)!}\left[\prod_{i=1}^{m-1}(s^2 - i^2)\right](\delta^{2m-1} f_{0+1/2} + \delta^{2m-1} f_{0-1/2})$$

$$+ \frac{s}{2 \times (2m)!}\left[\prod_{i=1}^{m-1}(s^2 - i^2)\right][(s-m) + (s+m)]\delta^{2m} f_0$$

$$(3.36)$$

This formula, being symmetric about x_0, is widely used in practice. The error term associated with Eq. (3.36) is exactly that of Eq. (3.35).

In the finite difference approximation, the central differences appearing in Eq. (3.36) must be expanded into linear combinations of f_i. Theoretically, this can be accomplished by successively lowering the order of the particular central difference according to Eq. (3.23), but it is a rather cumbersome procedure. Accordingly, we seek a more expedient representation of the central differences. The dual representations of the divided difference $f[x_{-m}, \ldots, x_0, \ldots, x_m]$ from Eqs. (3.1) and (3.32) allow us to write

$$\sum_{i=-m}^{m} f_i \left/ \prod_{\substack{j=-m \\ j \neq i}}^{m} (x_i - x_j) = \frac{\delta^{2m} f_0}{(2m)! h^{2m}}\right. \qquad (3.37)$$

To simplify this equation, note the following relation (see Appendix 3B):

$$\sum_{i=-m}^{m} \left[\prod_{\substack{j=-m \\ j \neq i}}^{m} (x_i - x_j) \right]$$

$$= \sum_{i=1}^{m} (-1)^{m+i}(h)^{2m}(m - i)!(m + i)!$$

$$+ \sum_{i=0}^{m} (-1)^{m-i}(h)^{2m}(m - i)!(m + i)! \tag{3.38}$$

It follows from Eqs. (3.37) and (3.38) that

$$\delta^{2m} f_0 = \sum_{i=1}^{m} (-1)^{m+i} \frac{(2m)!}{(m - i)!(m + i)!} f_{-i}$$

$$+ \sum_{i=0}^{m} (-1)^{m-i} \frac{(2m)!}{(m - i)!(m + i)!} f_i \tag{3.39}$$

In similar fashion, we further obtain

$$\delta^{2m-1} f_{0+1/2} = \sum_{i=1}^{m-1} (-1)^{m+i} \frac{(2m - 1)!}{(m - i - 1)!(m + i)!} f_{-i}$$

$$+ \sum_{i=0}^{m} (-1)^{m-i} \frac{(2m - 1)!}{(m - i)!(m + i - 1)!} f_i \tag{3.40}$$

$$\delta^{2m-1} f_{0-1/2} = \sum_{i=1}^{m} (-1)^{m+i-1} \frac{(2m - 1)!}{(m - i)!(m + i - 1)!} f_{-i}$$

$$+ \sum_{i=0}^{m-1} (-1)^{m-i-1} \frac{(2m - 1)!}{(m + i)!(m - i - 1)!} f_i \tag{3.41}$$

These formulas permit us to write any-order central differences directly into corresponding linear combinations of pivotal function values f_i.

3.3   APPROXIMATE EXPRESSIONS FOR DERIVATIVES

To replace the governing differential equations of elastic systems approximately by the finite difference equations, we represent derivatives

of the desired functions in terms of derivatives of corresponding interpolation polynomials. Depending on the desired location of the derivatives in the given interval of independent variable, we choose
appropriate interpolation polynomials developed in Section 3.2. Suppose that it is required to approximate the first and second derivatives
of $f(x)$ at one of the interior pivotal points, x_0, and express them in
terms of pivotal function values f_i near x_0. For this situation, naturally,
we select Stirling's formula with pivotal points arranged as shown in
Fig. 3.1c. On limiting Stirling's polynomial to second degree in x and
differentiating it twice with respect to x, we obtain at $x = x_0$

$$\frac{d}{dx} f_0 = \frac{d}{dx} P_2(x_0) + \frac{d}{dx} R_2(x_0)$$

$$\frac{d^2}{dx^2} f_0 = \frac{d^2}{dx^2} P_2(x_0) + \frac{d^2}{dx^2} R_2(x_0)$$

(3.42)

Since

$$\frac{d}{dx} = \frac{1}{h} \frac{d}{ds}$$

where $x = x_0 + hs$ and $dx = h \, ds$, we find from Eq. (3.36) that

$$\frac{d}{dx} P_2(x_0) = \frac{1}{2h} (\delta f_{0+1/2} + \delta f_{0-1/2}) = \frac{1}{2h} (-f_{-1} + f_1)$$

$$\frac{d^2}{dx^2} P_2(x_0) = \frac{1}{h^2} \delta^2 f_0 = \frac{1}{h^2} (f_{-1} - 2f_0 + f_1)$$

(3.43)

Suppressing the derivatives of the error term and introducing the approximation $F_i \approx f_i$, we arrived at the desired approximations

$$\frac{d}{dx} f_0 \approx \frac{1}{2h} (-F_{-1} + F_1), \qquad \frac{d^2}{dx^2} f_0 \approx \frac{1}{h^2} (F_{-1} - 2F_0 + F_1) \quad (3.44)$$

To estimate the error terms, we recall Eq. (3.20):

$$R_n^{(r)}(x) = \sum_{i=0}^{r} \frac{r!}{(n + r - i + 1)!i!} \, \omega_n^{(i)}(x) f^{(n+r-i+1)}(\xi_i) \qquad (3.45)$$

in which $\omega_n^{(i)}(x)$ for $n = 2$, $i = 0, 1, 2$, and $x = x_0$ are $\omega_2^{(0)}(x_0) = 0$, $\omega_2^{(1)}(x_0) = -h^2$, and $\omega_2^{(2)}(x_0) = 0$. Hence, Eq. (3.45) yields the estimates

$$\frac{d}{dx} R_2(x_0) = -\frac{h^2}{6} f^{(3)}(\xi_1)$$

$$\frac{d^2}{dx^2} R_2(x_0) = -\frac{h^2}{12} f^{(4)}(\xi_1) \qquad (3.46)$$

where ξ_1 is a point in the interval $x_{-1} < \xi_1 < x_1$. Since the approximations are carried out with the second-degree Stirling polynomial, we call these second-order approximations. Note that they entail three pivotal points, x_{-1}, x_0, and x_1. The associated errors are of $O(h^2)$.

Third- and fourth-order derivatives of $P_n(x)$ may be obtained by considering the fourth-degree Stirling polynomial $P_4(x)$. Derivatives of $P_4(x)$ at $x = x_0$ are

$$\frac{d}{dx} P_4(x_0) = \frac{1}{h} \left[\tfrac{1}{2}(\delta f_{0+1/2} + \delta f_{0-1/2}) - \tfrac{1}{12}(\delta^3 f_{0+1/2} + \delta^3 f_{0-1/2}) \right]$$

$$= \frac{1}{12h} (f_{-2} - 8f_{-1} + 8f_1 - f_2)$$

$$\frac{d^2}{dx^2} P_4(x_0) = \frac{1}{h^2} (\delta^2 f_0 - \tfrac{1}{12}\delta^4 f_0)$$

$$= \frac{1}{12h^2} (-f_2 + 16f_{-1} - 30f_0 + 16f_1 - f_2) \qquad (3.47)$$

$$\frac{d^3}{dx^3} P_4(x_0) = \frac{1}{h^3} \left[\tfrac{1}{2}(\delta^3 f_{0+1/2} + \delta^3 f_{0-1/2}) \right]$$

$$= \frac{1}{2h^3} (-f_{-2} + 2f_{-1} - 2f_1 + f_2)$$

$$\frac{d^4}{dx^4} P_4(x_0) = \frac{1}{h^4} \delta^4 f_0 = \frac{1}{h^4} (f_{-2} - 4f_{-1} + 6f_0 - 4f_1 + f_2)$$

The corresponding error terms become

$$\frac{d}{dx} R_4(x_0) = \frac{h^4}{30} f^{(5)}(\xi_1)$$

$$\frac{d^2}{dx^2} R_4(x_0) = \frac{h^4}{90} f^{(6)}(\xi_1)$$

$$\frac{d^3}{dx^3} R_4(x_0) = \frac{h^4}{210} f^{(7)}(\xi_1) - \frac{h^2}{4} f^{(5)}(\xi_2)$$

$$\frac{d^4}{dx^4} R_4(x_0) = \frac{h^4}{840} f^{(8)}(\xi_1) - \frac{h^2}{6} f^{(6)}(\xi_2)$$

(3.48)

where ξ_1 and ξ_2 are points in the interval containing x_{-2}, x_{-1}, x_0, x_1, and x_2 pivotal points. It may be seen that the fourth-order $dP_4(x_0)/dx$ and $d^2P_4(x_0)/dx^2$ are more accurate than the second-order $dP_2(x_0)/dx$ and $d^2P_2(x_0)/dx^2$, since their errors are of order $O(h^4)$ instead of $O(h^2)$. However, the expressions for $d^3P_4(x_0)/dx^3$ and $d^4P_4(x_0)/dx^4$ are accurate within $O(h^2)$. Higher-order finite difference formulas may be derived in a similar manner.

If derivatives are required at the endpoints of a given interval of x_i, the forward and backward Newton polynomials [Eqs. (3.26) and (3.27)] should be used. Tables 3.1, 3.2, and 3.3 show the second- and fourth-order approximations of $f^{(r)}(x_0)$ in forward, backward, and central finite differences, respectively, together with their error terms. On comparing Table 3.3 with Tables 3.1 and 3.2, we may conclude that in even-order derivatives, the central difference formulas possess errors that are smaller by the order $O(h)$ than those for forward and backward difference formulas. However, the order of errors are identical for all the odd-order derivatives.

Next, we derive finite difference formulas for unequal pivotal spacings, which may be used in the generation of finite difference equations near curved boundaries where unequal pivotal spacings are unavoidable. A detailed application is described in Section 3.4; at present, consider the second-order approximations for $f^{(r)}(x_0)$ in forward, backward, and central differences for the unequal pivotal spacings shown in Fig. 3.2. We refer to Newton's basic interpolation polynomial in terms of divided differences [Eq. (3.4)]. Differentiation of Eq. (3.4) gives at $x = x_0$

TABLE 3.1   Forward Difference Formulas[a]

Order of Approximation	Second-Order Forward Approximation, $n-2$	Fourth-Order Forward Approximation, $n-4$
Intervals (h)	0 1 2 →x x_0	0 1 2 3 4 →x x_0
$\dfrac{d}{dx} P_n(x_0)$	$\dfrac{1}{2h}(-3f_0 + 4f_1 - f_2)$	$\dfrac{1}{24h}(-50f_0 + 96f_1 - 72f_2 + 32f_3 - 6f_4)$
$\dfrac{d}{dx} R_n(x_0)$	$\dfrac{h^2}{3} f^{(3)}(\xi_1)$	$\dfrac{h^4}{5} f^{(5)}(\xi_1)$
$\dfrac{d^2}{dx^2} P_n(x_0)$	$\dfrac{1}{h^2}(f_0 - 2f_1 + f_2)$	$\dfrac{1}{12h^2}(35f_0 - 104f_1 + 114f_2 - 56f_3 + 11f_4)$
$\dfrac{d^2}{dx^2} R_n(x_0)$	$\dfrac{h^2}{6} f^{(4)}(\xi_1) - hf^{(3)}(\xi_2)$	$\dfrac{h^4}{15} f^{(6)}(\xi_1) - \dfrac{5h^3}{6} f^{(5)}(\xi_2)$
$\dfrac{d^3}{dx^3} P_n(x_0)$		$\dfrac{1}{4h^3}(-10f_0 + 36f_1 - 48f_2 + 28_3f - 6f_4)$
$\dfrac{d^3}{dx^3} R_n(x_0)$		$\dfrac{h^4}{35} f^{(7)}(\xi_1) - \dfrac{5h^3}{12} f^{(6)}(\xi_2) + \dfrac{7h^2}{4} f^{(5)}(\xi_3)$
$\dfrac{d^4}{dx^4} P_n(x_0)$		$\dfrac{1}{h^4}(f_0 - 4f_1 + 6f_2 - 4f_3 + f_4)$
$\dfrac{d^4}{dx^4} R_n(x_0)$		$\dfrac{h^4}{70} f^{(8)}(\xi_1) - \dfrac{5h^3}{21} f^{(7)}(\xi_2) + \dfrac{7h^2}{6} f^{(6)}(\xi_3)$ $- 2hf^{(5)}(\xi_4)$

[a] $\min(x_0, x_1, \ldots, x_n) < \xi_i < \max(x_0, x_1, \ldots, x_n)$.

$$\frac{d}{dx} f_0 = f[x_0, x_1] + (x_0 - x_1)f[x_0, x_1, x_2] + \frac{d}{dx} R_2(x_0)$$

(forward and backward)

$$= f[x_0, x_1] + (x_0 - x_1)f[x_1, x_0, x_1] + \frac{d}{dx} R_2(x_0) \qquad \text{(central)}$$

$$\frac{d^2}{dx^2} f_0 = 2f[x_0, x_1, x_2] + \frac{d^2}{dx^2} R_2(x_0) \qquad \text{(forward and backward)}$$

$$= 2f[x_{-1}, x_0, x_1] + \frac{d^2}{dx^2} R_2(x_0) \qquad \text{(central)}$$

(3.49)

Evaluating the divided differences in Eqs. (3.49) by means of Eq. (3.1) for the pivotal arrangements of Fig. 3.2, we obtain

TABLE 3.2 Backward Difference Formulas

Order of Approximation	Second-Order Backward Approximation, $n = 2$	Fourth-Order Backward Approximation, $n = 4$
Intervals (h)	$\begin{array}{ccc}-2 & -1 & 0\end{array}$ •——•——•——→ x x_0	$\begin{array}{ccccc}-4 & -3 & -2 & -1 & 0\end{array}$ •——•——•——•——•——→ x x_0
$\dfrac{d}{dx} P_n(x_0)$	$\dfrac{1}{2h}(f_{-2} - 4f_{-1} + 3f_0)$	$\dfrac{1}{24h}(6f_{-4} - 32f_{-3} + 72f_{-2} - 96f_{-1} + 50f_0)$
$\dfrac{d}{dx} R_n(x_0)$	$\dfrac{h^2}{3} f^{(3)}(\xi_1)$	$\dfrac{h^4}{5} f^{(5)}(\xi_1)$
$\dfrac{d^2}{dx^2} P_n(x_0)$	$\dfrac{1}{h^2}(f_{-2} - 2f_{-1} + f_0)$	$\dfrac{1}{12h^2}(11f_{-4} - 56f_{-3} + 114f_{-2} - 104f_{-1} + 35f_0)$
$\dfrac{d^2}{dx^2} R_n(x_0)$	$\dfrac{h^2}{6} f^{(4)}(\xi_1) - hf^{(3)}(\xi_2)$	$\dfrac{h^4}{15} f^{(6)}(\xi_1) + \dfrac{5h^3}{6} f^{(5)}(\xi_2)$
$\dfrac{d^3}{dx^3} P_n(x_0)$		$\dfrac{1}{4h^3}(6f_{-4} - 28f_{-3} + 48f_{-2} - 36f_{-1} + 10f_0)$
$\dfrac{d^3}{dx^3} R_n(x_0)$		$\dfrac{h^4}{35} f^{(7)}(\xi_1) + \dfrac{5h^2}{12} f^{(6)}(\xi_2) + \dfrac{7h^2}{4} f^{(5)}(\xi_3)$
$\dfrac{d^4}{dx^4} P_n(x_0)$		$\dfrac{1}{h^4}(f_{-4} - 4f_{-3} + 6f_{-2} - 4f_{-1} + f_0)$
$\dfrac{d^4}{dx^4} R_n(x_0)$		$\dfrac{h^4}{70} f^{(8)}(\xi_1) + \dfrac{5h^3}{21} f^{(7)}(\xi_1) + \dfrac{7h^2}{6} f^{(6)}(\xi_3) - 2hf^{(5)}(\xi_4)$

$$\frac{d}{dx} P_2(x_0) = \frac{1}{h}\left[-\frac{2\alpha + 1}{\alpha(\alpha + 1)} f_0 + \frac{\alpha + 1}{\alpha} f_1 - \frac{\alpha}{\alpha + 1} f_2 \right]$$

$$\frac{d^2}{dx^2} P_2(x_0) = \frac{2}{h^2}\left[\frac{1}{\alpha(\alpha + 1)} f_0 - \frac{1}{\alpha} f_1 + \frac{1}{(\alpha + 1)} f_2 \right] \quad \text{(forward)}$$

$$\frac{d}{dx} P_2(x_0) = \frac{1}{h}\left[\frac{\alpha}{(\alpha + 1)} f_{-2} - \frac{(\alpha + 1)}{\alpha} f_{-1} + \frac{2\alpha + 1}{\alpha(\alpha + 1)} f_0 \right]$$

$$\frac{d^2}{dx^2} P_2(x_0) = \frac{2}{h^2}\left[\frac{1}{(\alpha + 1)} f_{-2} - \frac{1}{\alpha} f_{-1} + \frac{1}{\alpha(\alpha + 1)} f_0 \right] \quad \text{(backward)}$$

$$\frac{d}{dx} P_2(x_0) = \frac{1}{h}\left[-\frac{\alpha}{(\alpha + \beta)\beta} f_{-1} + \frac{\beta - \alpha}{\alpha\beta} f_0 + \frac{\beta}{\alpha(\alpha + \beta)} f_1 \right]$$

$$\frac{d^2}{dx^2} P_2(x_0) = \frac{2}{h^2}\left[\frac{1}{(\alpha + \beta)\beta} f_{-1} - \frac{1}{\alpha\beta} f_0 + \frac{1}{(\alpha + \beta)\alpha} f_1 \right] \quad \text{(central)}$$

$$(3.50)$$

where $0 \leq \alpha \leq 1$ and $0 \leq \beta \leq 1$.

TABLE 3.3   Central Difference Formulas

Order of Approximation	Second-Order Central Approximation, $n = 2$	Fourth-Order Central Approximation, $n = 4$
Intervals (h)	$\begin{array}{ccc} -1 & 0 & 1 \\ \bullet & \bullet & \bullet \end{array} \longrightarrow x$ x_0	$\begin{array}{ccccc} -2 & -1 & 0 & 1 & 2 \\ \bullet & \bullet & \bullet & \bullet & \bullet \end{array} \longrightarrow x$ x_0
$\dfrac{d}{dx} P_n(x_0)$	$\dfrac{1}{2h}(-f_{-1} + f_1)$	$\dfrac{1}{12h}(f_{-2} - 8f_{-1} + 8f_1 - f_2)$
$\dfrac{d}{dx} R_n(x_0)$	$-\dfrac{h^2}{6} f^{(3)}(\xi_1)$	$\dfrac{h^4}{30} f^{(5)}(\xi_1)$
$\dfrac{d^2}{dx^2} P_n(x_0)$	$\dfrac{1}{h^2}(f_{-1} - 2f_0 + f_1)$	$\dfrac{1}{12h^2}(-f_{-2} + 16f_{-1} - 30f_0 + 16f_1 - f_2)$
$\dfrac{d^2}{dx^2} R_n(x_0)$	$-\dfrac{h^2}{12} f^{(4)}(\xi_1)$	$\dfrac{h^4}{90} f^{(6)}(\xi_1)$
$\dfrac{d^3}{dx^3} P_n(x_0)$		$\dfrac{1}{2h^3}(-f_{-2} + 2f_{-1} - 2f_1 + f_2)$
$\dfrac{d^3}{dx^3} R_n(x_0)$		$\dfrac{h^4}{210} f^{(7)}(\xi_1) - \dfrac{h^2}{4} f^{(5)}(\xi_3)$
$\dfrac{d^4}{dx^4} P_n(x_0)$		$\dfrac{1}{h^4}(f_{-2} - 4f_{-1} + 6f_0 - 4f_1 + f_2)$
$\dfrac{d^4}{dx^4} R_n(x_0)$		$\dfrac{h^4}{840} f^{(8)}(\xi_1) - \dfrac{h^2}{6} f^{(6)}(\xi_3)$

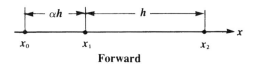

Forward

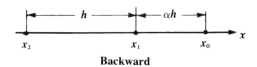

Backward

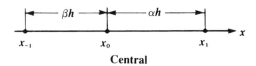

Central

Figure 3.2   Unequal pivotal spacings.

Application of Eq. (3.45) then yields the errors associated with Eq. (3.50). They are

$$\frac{d}{dx} R_2(x_0) = \frac{\alpha(\alpha + 1)h^2}{6} f^{(3)}(\xi_1)$$

(forward)

$$\frac{d^2}{dx^2} R_2(x_0) = \frac{\alpha(\alpha + 1)h^2}{12} f^{(4)}(\xi_1) - \frac{(2\alpha + 1)h}{3} f^{(3)}(\xi_2)$$

$$\frac{d}{dx} R_2(x_0) = \frac{\alpha(\alpha + 1)h^2}{6} f^{(3)}(\xi_1)$$

(backward)

$$\frac{d^2}{dx^2} R_2(x_0) = \frac{\alpha(\alpha + 1)h^2}{12} f^{(4)}(\xi_1) + \frac{(2\alpha + 1)h}{3} f^{(3)}(\xi_2)$$

$$\frac{d}{dx} R_2(x_0) = -\frac{\alpha\beta h^2}{6} f^{(3)}(\xi_1)$$

(central)

$$\frac{d^2}{dx^2} R_2(x_0) = -\frac{\alpha\beta h^2}{12} f^{(4)}(\xi_1) + \frac{(\beta - \alpha)h}{3} f^{(3)}(\xi_2)$$

(3.51)

When α and β are set equal to unity, Eqs. (3.50) and (3.51) reduce to formulas given in Tables 3.1, 3.2, and 3.3. From the error $d^2R_2(x_0)/dx^2$ for the central difference formula $d^2P_2(x_0)/dx^2$, note that the accuracy of $d^2P_2(x_0)/dx^2$ deteriorates with the use of unequal pivotal spacings.

3.4 TWO-DIMENSIONAL HARMONIC EQUATION, BIHARMONIC EQUATION, AND CURVED BOUNDARIES

In this section we consider finite difference approximations of partial derivatives that appear in the partial differential equations of elasticity. The method is analogous to the techniques developed for one independent variable. Without loss of generality, we restrict the treatment to problems of two independent variables (x, y). The extension to three or more independent variables is apparent.

For two-dimensional problems, an equally space rectangular mesh of the type shown in Fig. 3.3 is generally employed to define pivotal points as intersections of mesh lines. Thus, any pivotal point may be located by

$$x = x_0 + ih \quad \text{and} \quad y = y_0 + jk, \qquad i, j = 0, \pm 1, \pm 2, \ldots \quad (3.52)$$

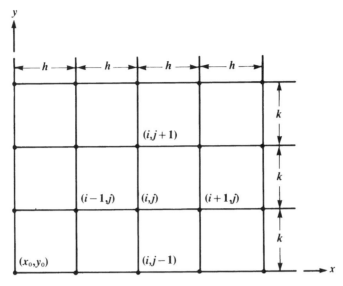

Figure 3.3 Rectangular mesh.

with respect to a conveniently chosen origin (x_0, y_0). The parameters h and k, denote the spacings of mesh lines in x and y directions, respectively. Whenever $h = k$, we say that the mesh is square.

Let $f(x, y)$ be a function of the independent variables (x, y). The function value of $f(x, y)$ at pivotal point $(x_0 + ih, y_0 + jk)$ is identified by appropriate subscripts. Thus, $f_{i,j}$ denotes $f(x_0 + ih, y_0 + jk)$. The first subscript is associated with the mesh line $x_0 + ih$, the second with the mesh line $y_0 + jk$.

Approximations to the derivatives $(\partial^r f / \partial x^r)_{i,j}$ and $(\partial^r f / \partial y^r)_{i,j}$ may be expressed by means of the forward, backward, or central difference formulas given in Tables 3.1, 3.2, and 3.3. On admitting errors of the order $O(h^2)$ or $O(k^2)$, for example, we write directly from Table 3.3 the second-order central finite difference formulas for partial derivatives:

$$\left(\frac{\partial f}{\partial x}\right)_{i,j} = \frac{1}{2h}\left(-f_{i-1,j} + f_{i+1,j}\right) + O(h^2)$$

$$\left(\frac{\partial^2 f}{\partial x^2}\right)_{i,j} = \frac{1}{h^2}\left(f_{i-1,j} - 2f_{i,j} + f_{i+1,j}\right) + O(h^2)$$

$$\left(\frac{\partial f}{\partial y}\right)_{i,j} = \frac{1}{2k}\left(-f_{i,j-1} + f_{i,j+1}\right) + O(k^2)$$

$$\left(\frac{\partial^2 f}{\partial y^2}\right)_{i,j} = \frac{1}{k^2}\left(f_{i,j-1} - 2f_{i,j} + f_{i,j+1}\right) + O(k^2)$$

(3.53)

To compute the mixed partial derivative $(\partial^2 f/\partial x \, \partial y)_{i,j}$, we first hold the y variable fixed and write

$$\left(\frac{\partial f}{\partial x}\right)_i = \frac{1}{2h}(-f_{i-1,j} + f_{i+1,j}) - \frac{h^2}{6}\left(\frac{\partial^3 f}{\partial x^3}\right)_{x=\xi} \tag{3.54}$$

where the single subscript refers to the x variable. Noting that every term of Eq. (3.54) is a function of y, we differentiate both sides with respect to y and evaluate the result at $y = y_0 + jk$. In particular, considering $(1/2h)(-f_{i-1} + f_{i+1})$ as a function of y, we apply $\partial/\partial y$ to obtain (see Table 3.3)

$$\left(\frac{\partial^2 f}{\partial x \partial y}\right)_{i,j}$$

$$= \frac{1}{(2h)(2k)}(f_{i-1,j-1} - f_{i-1,j+1} - f_{i+1,j-1} + f_{i+1,j+1}) \tag{3.55}$$

$$- \frac{k^2}{6}\left[\frac{\partial^3}{\partial y^3}\left(\frac{-f_{i-1} + f_{i+1}}{2h}\right)\right]_{y=\eta} - \frac{h^2}{6}\left(\frac{\partial^3 f}{\partial x^3}\right)_{\substack{x=\xi \\ y=y_0+jk}}$$

We transform the error term in parentheses by means of Eq. (3.54):

$$-\frac{k^2}{6}\frac{\partial^3}{\partial y^3}\left(\frac{-f_{i-1} + f_{i+1}}{2h}\right)_{y=\eta}$$

$$= -\frac{k^2}{6}\left\{\frac{\partial^3}{\partial y^3}\left[\left(\frac{\partial f}{\partial x}\right)_i + \frac{h^2}{6}\left(\frac{\partial^3 f}{\partial x^3}\right)_{x=\xi}\right]\right\}_{y=\eta}$$

$$= -\frac{k^2}{6}\left(\frac{\partial^4 f}{\partial x \partial y^3}\right)_{\substack{x=x_0+ih \\ y=\eta}} - \frac{h^2 k^2}{6}\left(\frac{\partial^6 f}{\partial x^3 \partial y^3}\right)_{\substack{x=\xi \\ y=\eta}}$$

where ξ and η are bounded by $x_0 + (i - 1)h < \xi < x_0 + (i + 1)h$ and $y_0 + (j - 1)k < \eta < y_0 + (j + 1)k$. Accordingly, we write

$$\left(\frac{\partial^2 f}{\partial x \, \partial y}\right)_{i,j}$$

$$= \frac{1}{4hk}(f_{i-1,j-1} - f_{i-1,j+1} - f_{i+1,j-1} + f_{i+1,j+1}) + O(h^2 \text{ or } k^2) \tag{3.56}$$

Analogously, differentiating the second equation of Eq. (3.53) with respect to y twice and estimating the error as above, we find that

$$
\left(\frac{\partial^4 f}{\partial x^2 \, \partial y^2} \right)_{i,j}
$$

$$
= \frac{1}{h^2 k^2} \, [f_{i-1,j-1} + f_{i-1,j+1} + f_{i+1,j-1} - f_{i+1,j+1} \tag{3.57}
$$

$$
- 2(f_{i-1,j} + f_{i,j+1} + f_{i+1,j} + f_{i,j-1}) + 4f_{i,j}] + O(h^2 \text{ or } k^2)
$$

If the fourth derivatives $(\partial^4 f / \partial x^4)_{i,j}$ and $(\partial^4 f / \partial y^4)_{i,j}$ are required, we employ the *fourth-order central difference formulas* of Table 3.3 to obtain

$$
\left(\frac{\partial^4 f}{\partial x^4} \right)_{i,j} = \frac{1}{h^4} \, (f_{i-2,j} - 4f_{i-1,j} + 6f_{i,j} - 4f_{i+1,j} + f_{i+2,j}) + O(h^2)
$$

$$
\tag{3.58}
$$

$$
\left(\frac{\partial^4 f}{\partial y^4} \right)_{i,j} = \frac{1}{k^4} \, (f_{i,j-2} - 4f_{i,j-1} + 6f_{i,j} - 4f_{i,j+1} + f_{i,j+2}) + O(k^2)
$$

Similarly, formulas for higher-order approximations for derivatives with respect to x, and for derivatives with respect to y, may be written.

Harmonic Equation

The plane harmonic equation (Laplace equation) is $\nabla^2 \phi = 0$, where $\phi = \phi(x, y)$ and ∇^2 denotes the Laplacian $\partial^2 / \partial x^2 + \partial^2 / \partial y^2$. By means of Eq. (3.53), the harmonic equation may be replaced by a finite difference equation at the pivotal point $(x_0 + ih, y_0 + jk)$ or simply (i, j); that is,

$$
(\nabla^2 \phi)_{i,j} = \frac{1}{h^2} \, (\phi_{i-1,j} - 2\phi_{i,j} + \phi_{i+1,j})
$$

$$
+ \frac{1}{k^2} \, (\phi_{i,j-1} - 2\phi_{i,j} + \phi_{i,j+1}) \tag{3.59}
$$

$$
+ O(h^2 \text{ or } k^2) = 0
$$

However, the representation of $\nabla^2 \phi = 0$ in Eq. (3.59) is not unique. For instance, by using fourth-order central formulas, we may write (see Table 3.3)

$$(\nabla^2 \phi)_{i,j} = \frac{1}{12h^2} (-\phi_{i-2,j} + 16\phi_{i-1,j} - 30\phi_{i,j}$$

$$+ \; 16\phi_{i+1,j} - \phi_{i+2,j})$$

$$+ \; \frac{1}{12k^2} (-\phi_{i,j-2} + 16\phi_{i,j-1} - 30\phi_{i,j}$$

$$+ \; 16\phi_{i,j+1} - \phi_{i,j+2}) + O(h^4 \text{ or } k^4) \tag{3.60}$$

Extension of these formulas to three variables (x, y, z) follows directly by assigning a third index and the associated spacing.

Biharmonic Equation

The plane biharmonic equation is $\nabla^4 \phi = \partial^4 \phi / \partial x^4 + 2\partial^4 \phi / \partial x^2 \, \partial y^2 + \partial^4 \phi / \partial y^4 = 0$. On replacing the partial derivatives occurring in this equation by Eqs. (3.57) and (3.58), we arrive at

$$(\nabla^4 \phi)_{i,j}$$

$$= \frac{1}{h^4} (\phi_{i-2,j} - 4\phi_{i-1,j} + 6\phi_{i,j} - 4\phi_{i+1,j} + \phi_{i+2,j})$$

$$+ \; \frac{2}{h^2 k^2} [\phi_{i-1,j-1} + \phi_{i-1,j+1}$$

$$+ \; \phi_{i+1,j-1} + \phi_{i+1,j+1} - 2(\phi_{i-1,j}$$

$$+ \; \phi_{i,j+1} + \phi_{i+1,j} + \phi_{i,j-1}) + 4\phi_{i,j}]$$

$$+ \; \frac{1}{k^4} (\phi_{i,j-2} - 4\phi_{i,j-1} + 6\phi_{i,j}$$

$$- \; 4\phi_{i,j+1} + \phi_{i,j+2}) + O(h^2 \text{ or } k^2) = 0 \tag{3.61}$$

For a square mesh $k = h$, Eq. (3.61) reduces to

$$(\nabla^4 \phi)_{i,j} = \frac{1}{h^4} [20\phi_{i,j} - 8(\phi_{i+1,j} + \phi_{i,j+1} + \phi_{i-1,j} + \phi_{i,j-1})$$

$$+ \; 2(\phi_{i+1,j+1} + \phi_{i-1,j+1} + \phi_{i-1,j-1} + \phi_{i+1,j-1})$$

$$+ \; (\phi_{i+2,j} + \phi_{i,j+2} + \phi_{i-2,j} + \phi_{i,j-2})] + O(h^2) = 0 \tag{3.62}$$

The corresponding formula of Eq. (3.62) with error of order $O(h^4)$ is

$$(\nabla^4 \phi)_{i,j} = \frac{1}{6h^4} [184\phi_{i,j} + 20(\phi_{i+1,j+1} + \phi_{i-1,j+1}$$

$$+ \phi_{i-1,j-1} + \phi_{i+1,j-1}) - 77(\phi_{i,j+1}$$

$$+ \phi_{i-1,j} + \phi_{i,j-1} + \phi_{i+1,j}) + 14(\phi_{i+2,j}$$

$$+ \phi_{i,j+2} + \phi_{i-2,j} + \phi_{i,j-2})$$

$$- (\phi_{i+3,j} + \phi_{i+2,j+1} + \phi_{i+1,j+2} + \phi_{i,j+3}$$

$$+ \phi_{i-1,j+2} + \phi_{i-2,j+1} + \phi_{i-3,j}$$

$$+ \phi_{i-2,j-1} + \phi_{i-1,j-2} + \phi_{i,j-3}$$

$$+ \phi_{i+1,j-2} + \phi_{i+2,j-1})] + O(h^4) = 0 \qquad (3.63)$$

Curved Boundaries

When the pivotal point $(x_0 + ih, y_0 + jk)$ is near a curved or irregular boundary as shown in Fig. 3.4, unequal mesh spacings may occur in

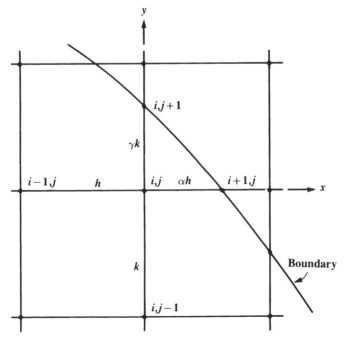

Figure 3.4 Unequal mesh near a curved boundary.

both x and y directions. To demonstrate techniques to deal with this situation, consider the approximation of the plane harmonic equation $\nabla^2 \phi = 0$ at such a pivotal point.

Basically, there are two types of bounary value problems associated with the equation, one with function values specified along the boundary (Dirichlet problem) and the other with normal derivative specified (Neumann problem). Accordingly, we present two versions of the finite difference approximations to $\nabla^2 \phi = 0$.

Consider first the Dirichlet problem where the ϕ value along the boundary is prescribed. Let the unequal mesh near the pivotal point (i, j) be defined as in Fig. 3.4, where $0 \leqslant \alpha \leqslant 1$ and $0 \leqslant \gamma \leqslant 1$. Applying the last formula of Eq. (3.50) to the function ϕ once with respect to x with $\beta = 1$ and $\alpha = \alpha$, and once with respect to y with $\beta = 1$ and $\alpha = \gamma$, we obtain

$$
\begin{aligned}
(\nabla^2 \phi)_{i,j} = \frac{2}{h^2} & \left(\frac{1}{(\alpha + 1)} \phi_{i-1,j} - \frac{1}{\alpha} \phi_{i,j} + \frac{1}{\alpha(\alpha + 1)} \phi_{i+1,j} \right) \\
+ \frac{2}{k^2} & \left(\frac{1}{(\gamma + 1)} \phi_{i,j-1} - \frac{1}{\gamma} \phi_{i,j} + \frac{1}{\gamma(\gamma + 1)} \phi_{i,j+1} \right) \\
+ O[(1 & - \alpha)h \text{ or } (1 - \gamma)k] = 0 \quad\quad\quad (3.64)
\end{aligned}
$$

where the magnitude of error is estimated by Eq. (3.61).

For the Neumann problem, consider the situation pictured in Fig. 3.5, where the regular pivotal points $(i, j + 1)$ and $(i + 1, j)$ fall outside the curved boundary. We construct two normals, n_1 and n_2, to the boundary through these two pivotal points. Let the intersections of these normals with the boundary be denoted by p and q and with the mesh lines, by r and s, as shown in the figure. The locations of r and s are defined by α and β parameters, respectively. For convenience, the distance between the pivotal point $(i, j + 1)$ and r is denoted by a, and the distance between the pivotal point $(i + 1, j)$ and s by b.

By means of the first-degree Newton interpolation polynomial of Eq. (3.4), we write

$$
\left(\frac{\partial \phi}{\partial n_1} \right)_p = \frac{1}{a} (\phi_{i,j+1} - \phi_r) \quad \text{and} \quad \left(\frac{\partial \phi}{\partial n_2} \right)_q = \frac{1}{b} (\phi_{i+1,j} - \phi_s)
$$

Solution for $\phi_{i,j+1}$ and $\phi_{i+1,j}$ from these equations yields

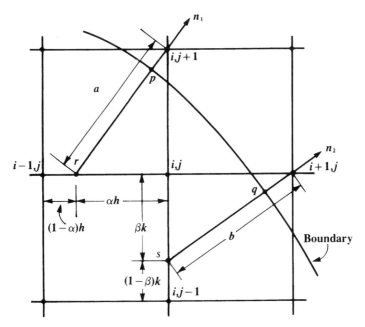

Figure 3.5 Unequal mesh near a curved boundary.

$$\phi_{i,j+1} = a \left(\frac{\partial \phi}{\partial n_1}\right)_P + \phi_r \qquad \phi_{i+1,j} = b \left(\frac{\partial \phi}{\partial n_2}\right)_q \qquad (3.65)$$

Repeated application of the first-degree Newton interpolation polynomial along jth and ith mesh lines, respectively, produces the following expressions for ϕ_r and ϕ_s:

$$\phi_r = \phi_{i,j} + (-\alpha h)\frac{\phi_{i-1,j} - \phi_{i,j}}{(-h)} = \alpha\phi_{i-1,j} + (1 - \alpha)\phi_{i,j}$$

$$\phi_s = \beta\phi_{i,j-1} + (1 - \beta)\phi_{i,j} \qquad (3.66)$$

Hence, from Eqs. (3.65) and (3.66), we have

$$\phi_{i,j+1} = a \left(\frac{\partial \phi}{\partial n_1}\right)_p + \alpha\phi_{i-1,j} + (1 - \alpha)\phi_{i,j}$$

$$\phi_{i+1,j} = b \left(\frac{\partial \phi}{\partial n_2}\right)_q + \beta\phi_{i,j-1} + (1 - \beta)\phi_{i,j} \qquad (3.67)$$

Next, elimination of $\phi_{i,j+1}$ and $\phi_{i+1,j}$ from Eq. (3.59) by the use of Eq. (3.67) leads to the desired approximate harmonic equation,

$$(\nabla^2 \phi)_{i,j} = \left[\frac{1}{h^2} \phi_{i-1,j} - (1 + \beta)\phi_{i,j} + \beta\phi_{i,j-1} + b \left(\frac{\partial \phi}{\partial n_2} \right)_q \right]$$

$$+ \frac{1}{k^2} \left[\phi_{i,j-1} - (1 + \alpha)\phi_{i,j} + \alpha\phi_{i-1,j} + a \left(\frac{\partial \phi}{\partial n_1} \right)_p \right]$$

$$+ O(h \text{ or } k) \tag{3.68}$$

Note that the use of the first-degree Newton interpolation polynomial leads to error of $O(h$ or $k)$.

3.5   FINITE DIFFERENCE APPROXIMATION OF THE PLANE STRESS PROBLEM

The equations of equilibrium for the plane stress problem of elasticity may be formulated in two ways: one in terms of the Airy stress function, which leads to the biharmonic partial differential equation, and the other in terms of (x, y) displacement components (u, v), which leads to a set of two coupled second-order partial differential equations (Boresi and Chong, 2000, Chap. 5). Since the boundary conditions associated with the Airy stress function formulation contain higher-order derivatives than those connected with the displacement formulation, in view of the nature of finite difference methods, the displacement formulation is preferred.

Accordingly, we consider the governing differential equation for the plane stress problem in the following form [Boresi and Chong, 2000, Eq. (5.8.3)]:

$$\frac{\partial^2 u}{\partial x^2} + \frac{1}{2}(1 - \nu)\frac{\partial^2 v}{\partial y^2} + \frac{1}{2}(1 + \nu)\frac{\partial^2 v}{\partial x\,\partial y} + \frac{1}{E}\left(\frac{\partial u}{\partial x} + \nu\frac{\partial v}{\partial y}\right)\frac{\partial E}{\partial x}$$

$$+ \frac{(1 - \nu)}{2E}\left(\frac{\partial u}{\partial y} + \frac{\partial v}{\partial x}\right)\frac{\partial E}{\partial y} = \frac{(1 + \nu)}{E}\frac{\partial}{\partial x}(EKT)$$

$$\frac{1}{2}(1 + \nu)\frac{\partial^2 u}{\partial x\,\partial y} + \frac{1}{2}(1 - \nu)\frac{\partial^2 v}{\partial x^2} + \frac{\partial^2 v}{\partial y^2} + \frac{(1 - \nu)}{2E}\left(\frac{\partial u}{\partial y} + \frac{\partial v}{\partial x}\right)\frac{\partial E}{\partial x}$$

$$+ \frac{1}{E}\left(\frac{\partial v}{\partial y} + \nu\frac{\partial u}{\partial x}\right)\frac{\partial E}{\partial y} = \frac{(1 + \nu)}{E}\frac{\partial}{\partial y}(EKT) \tag{3.69}$$

where $T = T(x, y)$ is a known temperature field and where the modulus of elasticity E, and the coefficient of linear thermal expansion K depend on (x, y). Poisson's ratio ν is taken to be constant.

The finite difference method may be applied to obtain an approximate solution to Eq. (3.69), subject to appropriate boundary conditions. For simplicity of demonstration, let $E =$ constant, although in general, this restriction is unnecessary. Then Eq. (3.69) reduces to

$$\frac{\partial^2 u}{\partial x^2} + \frac{1}{2}(1 - \nu)\frac{\partial^2 u}{\partial y^2} + \frac{1}{2}(1 + \nu)\frac{\partial^2 v}{\partial x\,\partial y} = (1 + \nu)\frac{\partial}{\partial x}(KT)$$

$$\frac{1}{2}(1 + \nu)\frac{\partial^2 u}{\partial x\,\partial y} + \frac{1}{2}(1 - \nu)\frac{\partial^2 v}{\partial x^2} + \frac{\partial^2 v}{\partial y^2} = (1 + \nu)\frac{\partial}{\partial y}(KT)$$

(3.70)

By way of illustration, consider the plane region $R: 0 \leqslant x \leqslant a, 0 \leqslant y \leqslant b$ (Fig. 3.6). Let the mesh dimensions be h and k. Let the (i, j) coordinates run from $(1$ to $m)$ and $(1$ to $n)$, respectively. Observe that there are mn pivotal points in R; hence, we have $2mn$ unknown pivotal u and v displacements.

To supply the $2mn$ equations required for the solution, we first consider the approximate satisfaction of the governing differential equation

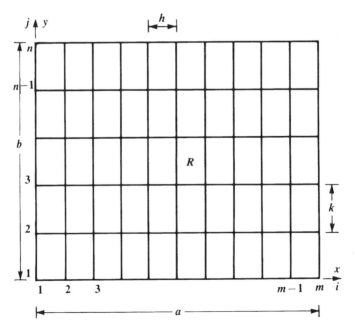

Figure 3.6   A rectangular region *R*.

within the region R. Thus, at the $2(m - 2)(n - 2)$ interior pivotal points, we require that the finite difference equations that replace Eq. (3.70) be satisfied. Employing *second-order central difference formulas* (Table 3.3), we write the $2(m - 2)(n - 2)$ algebraic equations that approximate Eq. (3.70):

$$\frac{1}{h^2} (U_{i+1,j} - 2U_{i,j} + U_{i-1,j}) + \frac{1 - \nu}{2k^2} (U_{i,j+1} - 2U_{i,j} + U_{i,j-1})$$

$$+ \frac{1 + \nu}{8hk} (V_{i+1,j+1} - V_{i+1,j-1} - V_{i-1,j+1} + V_{i-1,j-1})$$

$$- \frac{1 + \nu}{2h} (KT_{i+1,j} - KT_{i-1,j}) = 0$$

$$\frac{1 + \nu}{8hk} (U_{i+1,j+1} - U_{i+1,j-1} - U_{i-1,j+1} - U_{i-1,j-1})$$

$$+ \frac{1 - \nu}{2h^2} (V_{i+1,j} - 2V_{i,j} + V_{i-1,j}) + \frac{1}{k^2} (V_{i,j+1} - 2V_{i,j} + V_{i,j-1})$$

$$- \frac{1 + \nu}{2k} (KT_{i,j+1} - KT_{i,j-1}) = 0 \tag{3.71}$$

for $i = 2, 3, \ldots, m - 1$ and $j = 2, 3, \ldots, n - 1$. Here, (U, V) are approximations to (u, v).

Next, we derive the remaining required $4m + 4n - 8$ equations from consideration of boundary conditions. A variety of boundary conditions exist for the plane stress problem. However, the simplest one, in the sense of the present numerical scheme, is the case where the edges are prevented from displacing ($u = v = 0$ on the edges). In this case, we simply assign u and v values along the $2m + 2n - 4$ boundary pivotal points to obtain another $4m + 4n - 8$ equations.

Alternatively, we may consider the free edge boundary conditions where the boundary edges are free of any stress tractions. In this case, we must eliminate rigid-body motion (Boresi and Chong, 2000, Chap. 2). The translatory and rotatory rigid-body displacements of the region R can be prevented if the following conditions are satisfied:

$$U_{1,1} = V_{1,1} = V_{2,1} = 0 \tag{3.72}$$

In general, the stress-displacement relations for plane stress are [Boresi and Chong, 2000, Eq. (5.8.2)]

$$\sigma_x = \frac{E}{1 - \nu^2} \left[\frac{\partial u}{\partial x} + \nu \frac{\partial u}{\partial y} - (1 + \nu)KT \right]$$

$$\sigma_y = \frac{E}{1 - \nu^2} \left[\frac{\partial v}{\partial y} + \nu \frac{\partial u}{\partial x} - (1 + \nu)KT \right] \qquad (3.73)$$

$$\tau_{xy} = G \left(\frac{\partial u}{\partial y} + \frac{\partial v}{\partial x} \right); \qquad G = \frac{E}{2(1 + \nu)}$$

For the edge $y = 0$ free, $\sigma_y = \tau_{xy} = 0$. Hence, Eq. (3.73) yield for $y = 0$

$$\frac{\partial v}{\partial y} + \nu \frac{\partial u}{\partial x} - (1 + \nu)KT = 0$$

$$(3.74)$$

$$\frac{\partial u}{\partial y} + \frac{\partial v}{\partial x} = 0$$

Since the second-order central differences with error of $O(h^2$ or $k^2)$ have been employed in deriving Eq. (3.71), the same order of difference formulas should be used in replacing Eq. (3.74). Otherwise, the approximation may be inconsistent. Approximating Eq. (3.73), we use the following finite difference approximations of the derivatives of u and v along the edge $y = 0$:

$$\left(\frac{\partial v}{\partial y} \right)_{i,1} = \frac{1}{2k} (-3V_{i,1} + 4V_{i,2} - V_{i,3}) \qquad i = 3, \ldots, m$$

$$\left(\frac{\partial u}{\partial x} \right)_{i,1} = \frac{1}{2h} (-U_{i-1,1} + U_{i+1,1}) \qquad i = 2, \ldots, m - 1$$

$$\left(\frac{\partial u}{\partial x} \right)_{i,1} = \frac{1}{2h} (U_{i-2,1} - 4U_{i-1,1} + 3U_{i,1}) \qquad i = m$$

$$\left(\frac{\partial u}{\partial y} \right)_{i,1} = \frac{1}{2k} (-3U_{i,1} + 4U_{i,2} - U_{i,3}) \qquad i = 2, \ldots, m$$

$$\left(\frac{\partial v}{\partial x} \right)_{i,1} = \frac{1}{2h} (-V_{i-1,1} + V_{i+1,1}) \qquad i = 2, \ldots, m - 1$$

$$\left(\frac{\partial v}{\partial x} \right)_{i,1} = \frac{1}{2h} (V_{i-2,1} - 4V_{i-1,1} + 3V_{i,1}) \qquad i = m$$

where forward, backward, and central difference formulas of Tables

3.1, 3.2, and 3.3 are used. These expressions permit us to approximate Eq. (3.74) as

$$\frac{1}{2k}(-3V_{i,1} + 4V_{i,2} - V_{i,3}) + \frac{\nu}{2h}(-U_{i-1,1} + U_{i+1,1})$$

$$-(1 + \nu)KT_{i,1} = 0 \qquad i = 3, \ldots, m - 1$$

$$\frac{1}{2k}(-3U_{i,1} + 4U_{i,2} - U_{i,3}) + \frac{1}{2h}(-V_{i-1,1} + V_{i+1,1}) = 0$$

$$i = 2, \ldots, m - 1$$

$$\frac{1}{2k}(-3V_{i,1} + 4V_{i,2} - V_{i,3}) + \frac{\nu}{2h}(U_{i-2,1} - 4U_{i-1,1} + 3U_{i,1})$$

$$-(1 + \nu)KT_{i,1} = 0 \qquad i = m$$

$$\frac{1}{2k}(-3U_{i,1} + 4U_{i,2} - U_{i,3}) + \frac{1}{2h}(V_{i-2,1} - 4V_{i-1,1} + 3V_{i,1}) = 0$$

$$i = m \quad (3.75)$$

Equation (3.75) represents $(m - 2) + (m - 1)$ algebraic equations.

Along a free edge $y = b$, again $\sigma_y = \tau_{xy} = 0$. Hence, in an analogous manner, we write another $2(m - 1)$ equations:

$$\frac{1}{2k}(V_{i,n-2} - 4V_{i,n-1} + 3V_{i,n})$$

$$+ \frac{\nu}{2h}(-U_{i-1,n} + U_{i+1,n}) - (1 + \nu)KT_{i,n} = 0$$

$$i = 2, \ldots, m - 1$$

$$\frac{1}{2k}(U_{i,n-2} - 4U_{i,n-1} + 3U_{i,n})$$

$$+ \frac{1}{2h}(-V_{i-1,n} + V_{i+1,n}) = 0 \qquad i = 2, \ldots, m - 1$$

$$\frac{1}{2k}(V_{i,n-2} - 4V_{i,n-1} + 3V_{i,n})$$

$$+ \frac{\nu}{2h}(U_{i-2,n} - 4U_{i-1,n} + 3U_{i,n}) - (1 + \nu)KT_{i,n} = 0 \qquad i = m$$

$$\frac{1}{2k} (U_{i,n-2} - 4U_{i,n-1} + 3U_{i,n})$$

$$+ \frac{1}{2h} (V_{i-2,n} - 4V_{i-1,n} + 3V_{i,n}) = 0 \qquad i = m$$

$$(3.76)$$

Along a free edge $x = 0$, $\sigma_x = \tau_{xy} = 0$. Hence, by Eq. (3.73),

$$\frac{\partial u}{\partial x} + \nu \frac{\partial v}{\partial y} - (1 + \nu)KT = 0$$

$$(3.77)$$

$$\frac{\partial u}{\partial y} + \frac{\partial v}{\partial x} = 0$$

Equation (3.77) can be cast into corresponding finite difference equations in a manner similar to previous approximations. Upon writing the resulting difference equations at $(n - 1)$ pivotal points ($j = 2, 3, \ldots, n$) along the free edge $x = 0$, and $(n - 2)$ pivotal points ($j = 2, 3, \ldots, n - 1$) along the free edge $x = a$, we further obtain $2(n - 1) + 2(n - 2)$ equations for the case of a rectangular region with free edges.

Accordingly, the boundary conditions provide a total of $4m + 4n - 8$ equations. Equation (3.71) combined with these boundary equations form $2mn$ linear algebraic equations for the $2mn$ unknowns ($U_{i,j}$, $V_{i,j}$). The solution of these equations subject to Eq. (3.72) represents an approximate solution of the plane stress problem of elasticity for the rectangular region with stress free edges. Various methods of solving such equations in conjunction with digital computers have been developed (Collatz, 1960; Shoup, 1979; Chapra, 1990).

3.6 TORSION PROBLEM

In Chapter 7 of Boresi and Chong (2000) we formulated the torsion of elastic prismatic bars in terms of Saint-Venant's warping function and also in terms of Prandtl's torsion function. Prandtl's formulation leads to a simpler Dirichlet-type boundary value problem [Boresi and Chong, 2000, Eq. (7.3.16)]:

$$\nabla^2\phi = -2G\beta \qquad \text{over region } R$$
$$\phi = 0 \qquad \text{on contour } C$$

(3.78)

where C is the bounding curve of the bar cross section R and β is the angle of twist per unit length of the bar. Accordingly, we employ Eq. (3.78) in the finite difference analysis of the torsion problem.

Square Cross Section

Let the bar cross section R be square (Fig. 3.7). Then, by symmetry, we need consider only a quarter of the region R. Figure 3.7 shows a 4×4 square mesh subdivision with mesh width $h = a/4$. Numbering the pivotal points as shown in Fig. 3.7, we need consider only the three values $F_{1,1}$, $F_{2,1}$, and $F_{2,2}$, where $F_{i,j}$ are approximations to $\phi_{i,j}$. Note that the $\phi_{i,j}$ on the contour C are identically zero.

Employing the second-order central difference formula for $\nabla^2\phi$ [Eq. (3.59)], we write the governing differential equation in finite difference

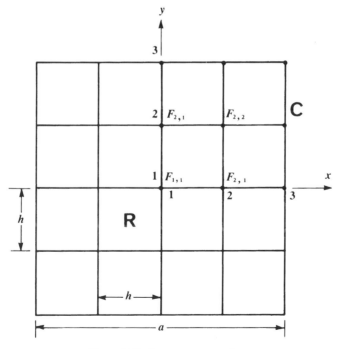

Figure 3.7 Square cross section.

form for the three interior pivotal points to obtain (since $\phi_{i,j} = 0$ on the boundary)

$$-4F_{1,1} + 4F_{2,1} = -2G\beta h^2$$

$$F_{1,1} - 4F_{2,1} + 2F_{2,2} = -2G\beta h^2 \qquad (3.79)$$

$$2F_{2,1} - 4F_{2,2} = -2G\beta h^2$$

Solution of Eq. (3.79) yields

$$F_{1,1} = 2.250G\beta h^2$$

$$F_{2,1} = 1.750G\beta h^2 \qquad (3.80)$$

$$F_{2,2} = 1.375G\beta h^2$$

Next, we determine the maximum shearing stress $\tau_{zy} = \partial\phi/\partial x$ that occurs at the point $(x, y) = (a/2, 0)$. On taking account of the symmetry and using fourth-order backward difference formulas (Tables 3.2), we approximate $\partial\phi/\partial x$ at $(x, y) = (a/2, 0)$ as

$$\left(\frac{\partial\phi}{\partial x}\right)_{\substack{x=a/2 \\ y=0}} = \frac{1}{24h}(0 - 32F_{2,1} + 72F_{1,1} - 96F_{2,1} + 0) \quad (3.81)$$

Substitution of Eq. (3.80) into (3.81) gives

$$(\tau_{zy})_{max} = -2.583G\beta h = -0.646G\beta a$$

The exact solution (Boresi and Chong, 2000, Sec. 7.10) yields $(\tau_{zy})_{max} = -0.675G\beta a$. If we take a finer mesh, we obtain a better approximation. For example, with $h = a/8$, we find that $(\tau_{zy})_{max} = -0.666G\beta a$.

Bar with Elliptical Cross Section

An explicit closed-form analytical solution of the torsion problem with curved or irregular boundary contours generally is not available, except for a few special cases. However, approximate solutions by the finite difference method are still feasible, although generation of the approximating finite difference equations becomes numerically more involved.[2] To illustrate the numerical procedure, we consider torsion of a bar with elliptical cross section for which an exact solution is available (Boresi and Chong, 2000, Sec. 7.4). By comparison with the exact solution, we may verify the accuracy of the finite difference approximate solution.

By way of illustration, let us subdivide the elliptical cross section ($a = 10$ and $b = 7$), as shown in Fig. 3.8. Only 11 unknown interior pivotal function values F_i are required because of the symmetric nature of the problem. For simplicity we have denoted each pivotal point by a single number. Hence, F_i represents the approximate value of the Prandtl torsion function ϕ at the pivotal point i (Fig. 3.8). We see from Fig. 3.8 that the finite difference approximation [Eq. (3.59)] that replaces $\nabla^2 \phi = -2G\beta$ may be written at interior pivotal points 1, 2, ..., 7 with constant mesh interval of width $h = 2.5$. However, unequal pivotal spacings occur adjacent to pivotal points 8, 9, 10, and 11 near the boundary. Therefore, for these pivotal points Eq. (3.64) is used to approximate $\nabla^2 \phi = -2G\beta$. In conjunction with the application of Eq. (3.64), the unequal pivotal spacings adjacent to points 8, 9, 10, and 11

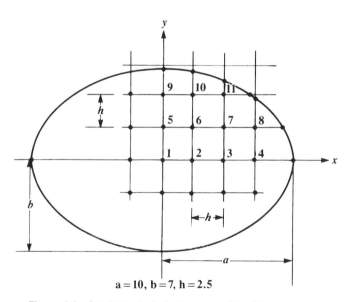

$a = 10, b = 7, h = 2.5$

Figure 3.8 Subdivided elliptical cross section (11 unknowns).

[2]It is shown in Chapter 4 that the inherent difficulties of the finite difference method in dealing with curved boundaries or in satisfying complicated boundary conditions can be circumvented by use of the finite element method.

are shown in Fig. 3.9. Accordingly, by Eqs. (3.59) and (3.64), we obtain 11 algebraic equations for the 11 unknowns, F_i. They are

$$-4F_1 + 2F_2 + 2F_5 = -2G\beta h^2$$

$$F_1 - 4F_2 + F_3 + 2F_6 = -2G\beta h^2$$

$$F_2 - 4F_3 + F_4 + 2F_7 = -2G\beta h^2$$

$$F_3 - 4F_4 + F_8 = -2G\beta h^2$$

$$F_1 - 4F_5 + 2F_6 + F_9 = -2G\beta h^2$$

$$F_2 + F_5 - 4F_6 + F_7 + F_{10} = -2G\beta h^2 \qquad (3.82)$$

$$F_3 + F_6 - 4F_7 + F_8 + F_{11} = -2G\beta h^2$$

$$1.1111F_5 - 4.5000F_9 + 2F_{10} = -2G\beta h^2$$

$$1.1688F_6 + F_9 - 4.8126F_{10} + F_{11} = -2G\beta h^2$$

$$1.4036F_7 + 1.1114F_{10} - 7.2090F_{11} = -2G\beta h^2$$

$$1.0799F_4 + 1.1519F_7 - 5.0639F_8 = -2G\beta h^2$$

where the condition $\phi = 0$ on the boundary has been used.

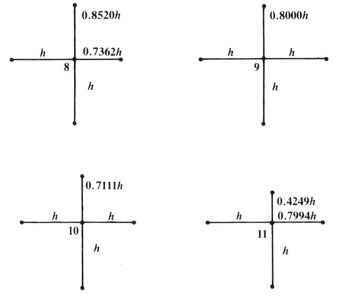

Figure 3.9 Unequal pivotal spacings ($h = 2.5$).

Solution of Eq. (3.82) yields the 11 unknowns, F_i. The F_i values along the y axis, designated as approximate ϕ values, are listed in Table 3.4 for $h = 2.5$ and $h = 1.5$, along with corresponding exact values. From the F_i values for $h = 2.5$, we calculate $(\tau_{zx})_{max} = \partial\phi/\partial y$ at $(x, y) = (0, 7)$ by means of the second-degree Newton interpolation formula of Eq. (3.4). The approximate value $(\tau_{zx})_{max} = -9.0475G\beta$ is 3.7% less than the corresponding exact value $(\tau_{zx})_{max} = -9.3960G\beta$.

For $h = 1.5$ we obtain a finer mesh subdivision with 30 unknown F_i (Fig. 3.10). The approximate maximum shearing stress $(\tau_{zx})_{max} = -9.3702G\beta$ differs by only 0.28% from the exact value (Table 3.4).

We note that the torsional function ϕ can be approximated fairly accurately even with a rough mesh subdivision. However, since the determination of shearing stress requires numerical differentiation of the approximate torsion function, for a given mesh size, the approximation of the shear stress is less accurate than that of the stress function (Table 3.4). Accordingly, in general, accurate approximation of the shear stress requires a finer mesh subdivision than does accurate approximation of the torsion function.

APPENDIX 3A: DERIVATION OF EQ. (3.16)

By Eq. (3.14) with $p = 1$ and $n = 1$, we have with Eq. (3.1),

$$g[b_0, b_1] = \frac{g(b_1) - g(b_0)}{b_1 - b_0} \tag{3A.1}$$

Let $b_0 = x$ and $b_1 = x + \varepsilon$. Then Eq. (3A.1) becomes

TABLE 3.4   Comparison of Exact and Approximate Values of Torsion Function and Maximum Shearing Stress

	11 Unknowns with $h = 2.5$			30 Unknowns with $h = 1.5$	
y	Exact $\phi/G\beta h^2$	Approximate $\phi/G\beta h^2$	y	Exact $\phi/G\beta h^2$	Approximate $\phi/G\beta h^2$
0.0	5.2617	5.2616	0.0	14.6159	14.5861
2.5	4.5906	4.5905	1.5	13.9448	13.9155
5.0	2.5772	2.5771	3.0	11.9314	11.9044
			4.5	8.5756	8.5542
			6.0	3.8777	3.8673
	Exact	Approximate		Exact	Approximate
$(\tau_{zx})_{max}$	$-9.3960G\beta$	$-9.0475G\beta$		$-9.3960G\beta$	$-9.3702G\beta$

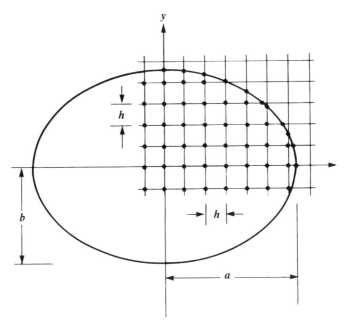

Figure 3.10   Subdivided elliptical cross section (30 unknowns).

$$g[x, x + \varepsilon] = \frac{g(x + \varepsilon) - g(x)}{\varepsilon} \tag{3A.2}$$

Consider the limit of Eq. (3A.2) as $\varepsilon \to 0$. Then

$$g[x, x + \varepsilon] \to g[x, x] = \lim_{\varepsilon \to 0} \frac{g(x + \varepsilon) - g(x)}{\varepsilon} = g'(x) \quad \text{(3A.3)}$$

where the prime denotes derivative with respect to x. Thus, although $g[x, x]$ cannot be evaluated directly from the values of g and the definition of Eq. (3.1), it can be expressed in terms of the derivative of $g(x)$.

Similarly, consider

$$g[b_0, b_1, b_2] = \frac{g[b_1, b_2] - g[b_0, b_1]}{b_2 - b_0} \tag{3A.4}$$

Let $b_0 = x$, $b_1 = b_0 + \varepsilon$, $b_2 = b_1 + \varepsilon$, or $b_0 = x$, $b_1 = x + \varepsilon$, $b_2 = x + 2\varepsilon$. Then, Eq. (3A.4) may be written

$$g[b_0, b_1, b_2] = g[x, x + \varepsilon, x + 2\varepsilon]$$

$$= \frac{g[x + \varepsilon, x + 2\varepsilon] - g[x, x + \varepsilon]}{2\varepsilon}$$

$$= \frac{g[x + 2\varepsilon] - g[x + \varepsilon]}{2\varepsilon \cdot \varepsilon} - \frac{g[x + \varepsilon] - g[x]}{2\varepsilon \cdot \varepsilon}$$

$$= \frac{g[x + 2\varepsilon] - 2g[x + \varepsilon] + g[x]}{2\varepsilon^2}$$

Taking the limit as $\varepsilon \to 0$, we have

$$g[b_0, b_1, b_2] \xrightarrow[\varepsilon \to 0]{} g[x, x, x] = \lim_{\varepsilon \to 0} \frac{g[x + 2\varepsilon] - 2g[x + \varepsilon] + g[x]}{2\varepsilon^2}$$

$$\to \frac{1}{2!} g''(x)$$

Continuing in this manner, we obtain

$$g[x, x, \ldots, x] = \frac{1}{n!} g^n(x) \tag{3A.5}$$

where x is repeated $n + 1$ times. [See Eq. (3.16).]

APPENDIX 3B: DERIVATION OF EQ. (3.38)

Since $x_i = x_0 + ih$ and $x_j = x_0 + jh$, we have $x_i - x_j = h(i \quad j)$. Now considering the left-hand side of Eq. (3.38), we expand in the following manner:

$$\sum_{i=-m}^{m} \left[\prod_{\substack{j=-m \\ j \neq i}}^{m} (x_i - x_j) \right] = \sum_{i=0}^{m} \left[\prod_{j=-1}^{-m} (x_i - x_j) \prod_{\substack{j=0 \\ j \neq i}}^{m} (x_i - x_j) \right]$$

$$+ \sum_{i=-1}^{-m} \left[\prod_{j=-1}^{-m} (x_i - x_j) \prod_{j=0}^{m} (x_i - x_j) \right] \tag{3B.1}$$

With the condition $x_i - x_j = h(i - j)$, the right-hand side of Eq. (3B.1) can be written in the form

$$\sum_{i=0}^{m} [h^m(i+1)(i+2) \cdots (i+m)h^m(i)$$

$$\times (i-1) \cdots (i-j-1)(i-j+1) \cdots (i-m)]$$

$$+ \sum_{i=-1}^{-m} [h^m(i+1)(i+2) \cdots (-1)(1) \cdots (i$$

$$+ m)h^m(i)(i-1) \cdots (i-m)] \tag{3B.2}$$

Collecting terms, we find that Eq. (3B.2) can be written in the form

$$\sum_{i=0}^{m} h^{2m}(-1)^{m-i}(m+i)!(m-i)! + \sum_{i=1}^{m} h^{2m}(-1)^{m+i}(m+i)!(m-i)!$$
$$\tag{3B.3}$$

where in the last summation we have taken $i = -i$ and $m = -m$ in order to sum with positive values of i. Substitution of Eq. (3B.3) for the right-hand side of Eq. (3B.1) yields Eq. (3.38).

REFERENCES

Boresi, A. P., and Chong, K. P. 2000. *Elasticity in Engineering Mechanics,* Wiley, New York.

Bushnell, D. 1973. Finite difference energy models versus finite element models: two variational approaches in one computer program, *ONR Symposium,* Academic Press, San Diego, CA, pp. 291–336.

Chapra, S. C., 1990. *Numerical Methods for Engineers with Personal Computer Applications,* McGraw-Hill, New York.

Collatz, L. 1960. *The Numerical Treatment of Differential Equations,* 3rd ed., Springer-Verlag, Berlin.

Isaacson, E., and Keller, H. B. 1966. *Analysis of Numerical Methods,* Wiley, New York.

Shoup, T. E. 1979. *A Practical Guide to Computer Methods for Engineers,* Prentice-Hall, Upper Saddle River, NJ.

Stanton, R. G. 1961. *Numerical Methods for Science and Engineering,* Prentice-Hall, Upper Saddle River, NJ.

Thomas, G. B., Jr. 1960. *Calculus and Analytical Geometry,* 3rd ed., Addison-Wesley, Reading, MA, pp. 134–135.

Zienkiewicz, O. C., and Morgan, K. 1983. *Finite Elements and Approximation,* Wiley, New York.

BIBLIOGRAPHY

Allaire, P. E. 1985. *Basics of the Finite Element Method,* Wm. C. Brown, Dubuque, IA, Chap. 16.

Bertsekas, D. P., and Tsitsiklis, J. 1997. *Parallel and Distributed Computation: Numeerical Methods,* Athena Scientific, Belmont, MA.

Boresi, A. P., and Chong, K. P. (eds.) 1984. *Engineering Mechanics in Civil Engineering,* Vols. 1 and 2, American Society of Civil Engineers (Engineering Mechanics Division), Reston, VA.

Chong, K. P., and Miller, M. M., Jr. 1979. Behavior of thin hyperbolic shells on rectangular enclosures, in *Proc. IASS World Congress on Shell and Spatial Structures,* IASS, Madrid, pp. 421–435.

Edwards, B. 1999. *Calculus with Analytic Geometry,* Prentice Hall, Upper Saddle River, NJ.

Fillerup, J. M., and Boresi, A. P. 1982. Excitation of finite viscoelastic solid on springs, *Nucl. Eng. Des.,* **71,** 179–193.

Hartree, D. R. 1961. *Numerical Analysis,* 2nd ed., Oxford University Press, Oxford.

Kikuchi, N. 1986. *Finite Element Methods in Mechanics,* Cambridge University Press, Cambridge.

La Fara, R. L. 1973. *Computer Methods for Science and Engineering,* Hayden, Rochelle Park, NJ.

Langhaar, H. L. 1969. Two methods that converge to the method of least squares, *J. Franklin Inst.,* **288,** 165–173.

Langhaar, H. L., and Chu, S. C. 1970. Piecewise polynomials and the partition method for ordinary differential equations, in *Developments in Theoretical and Applied Mechanics,* Vol. 8, Frederick, D. (ed.), Pergamon Press, New York, pp. 553–564.

Langhaar, H. L., Boresi, A. P., and Miller, R. E. 1974. Periodic excitation of a finite linear viscoelastic solid, *Nucl. Eng. Des.,* **30,** 349–368.

Lapidus, L., and Pinder, G. F. 1982. *Numerical Solution of Partial Differential Equations in Science and Engineering,* Wiley, New York.

Mitchell, A. R., and Griffiths, D. F. 1980. *The Finite Difference Method in Partial Differential Equations,* Wiley, New York.

Ralston, A., and Rabinowitz, P. 2001. *A First Course in Numerical Analysis,* 2nd ed., Dover Publications, Mineola, NY.

Sharma, S. K., and Boresi, A. P. 1978. Finite element weighted residual methods: axisymmetric shells, *J. Eng. Mech. Div. ASCE,* **104**(EM4), 895–909.

Taflove, A., and Hagness, S. C. 2000. *Computational Electrodynamics: The Finite-Difference Time-Domain Method,* 2nd ed., Artech House, Dedham, MA.

Vandergraft, J. S. 1978. *Introduction to Numerical Computations,* Academic Press, San Diego, CA.

Williams, P. W. 1972. *Numerical Computation,* Harper & Row, New York.

Zienkiewicz, O. C. Chan, A. H. C. Pastor, M., and Schrefler, B. A. 1999. *Computational Geomechanics: with Special Reference to Earthquake Engineering,* Wiley, New York.

4

The Finite Element Method

4.1 INTRODUCTION

The finite element method,[1] the most powerful numerical technique available today for the analysis of complex structural and mechanical systems, can be used to obtain numerical solutions to a wide range of problems. The method can be used to analyze both linear and nonlinear systems. Nonlinear analysis can include material yielding, creep, or cracking; aeroelastic response; buckling and postbuckling response; and contact and friction. It can be used for both static and dynamic analysis. In its most general form, the finite element method is not restricted to structural (or mechanical) systems. It has been applied to problems in fluid flow, heat transfer, and electric potential. Its versatility is a major reason for the popularity of the method.

A complete study of finite element methods is beyond the scope of this book. So the objective of this chapter is to outline the basic formulation for problems in linear elasticity. The formulation for plane elasticity is presented first. Then use of the method to analyze framed structures is examined. Finally, accuracy, convergence, and proper modeling techniques are discussed. Advanced topics such as analysis of

---

[1] The discovery of the method is often attributed to Courant (1943). The use of the method in structural (aircraft) analysis was first reported by Turner et al. (1956). The method received its name from Clough (1960).

plate bending and shell problems, three-dimensional problems, and dynamic and nonlinear analysis, are left to more specialized texts.

Analytical Perspective

The classical method of analysis in elasticity involves the study of an infinitesimal element of an elastic body (continuum or domain). Relationships among stress, strain, and displacement for the infinitesimal element are developed (Boresi and Chong, 2000) that are usually in the form of differential (or integral) equations that apply to each point in the body. These equations must be solved subject to appropriate boundary conditions. In other words, the approach is to define and solve a classical boundary value problem in mathematics (Boresi and Chong, 2000). Problems in engineering usually involve very complex shapes and boundary conditions. Consequently, for such cases, the equations cannot be solved exactly but must finally be solved by approximate methods: for example, by truncated series, finite differences, numerical integration, and so on. All of these approximate methods require some form of discretization of the solution.

By contrast, the formulation of finite element solutions recognizes at the outset that discretization is likely to be required. The first step in application of the method is to discretize the domain into an assemblage of a finite number of finite-size *elements* (or subregions) that are connected at specified node points. The quantities of interest (usually just displacements) are assumed to vary in a particular fashion over the element. This *assumed* element behavior leads to relatively simple integral equations for the individual elements. The integral equations for an element are evaluated to produce algebraic equations (in the case of static loading) in terms of the displacements of the node points. The algebraic equations for all elements are assembled to achieve a system of equations for the structure as a whole. Appropriate numerical methods are then used to solve this system of equations.

In summary, using the classical approach, we often are confronted with partial differential (or integral) equations that cannot be solved in closed form. This is due to the complexity of the geometry of the domain or of the boundary conditions. Consequently, we are forced to use numerical methods to obtain an approximate solution. These numerical methods always involve some type of discretization. In the finite element method, the discretization is performed at the outset. Then further approximation in either the formulation or the solution may not be necessary.

Sources of Error

There are three sources of error in the finite element method: errors due to approximation of the domain (discretization error), errors due to approximation of the element behavior (formulation error), and errors due to use of finite-precision arithmetic (numerical error).

Discretization error is due to the approximation of the domain with a finite number of elements of fixed geometry. For instance, consider the analysis of a rectangular plate with a centrally located hole, Fig. 4.1*a*. Due to symmetry, it is sufficient to model only one-fourth of the plate. If the region is subdivided into triangular elements (a triangular mesh or grid), the circular hole is approximated by a series of straight lines. If a few large triangles are used in a coarse mesh (Fig. 4.1*b*), a greater discretization error results than if a large number of small elements is used in a fine mesh (Fig. 4.1*c*). Other geometric shapes may be chosen. For example, with quadrilateral elements that can represent curved sides, the circular hole is more accurately approximated (Fig. 4.1*d*). Hence, discretization error may be reduced by grid refinement. The grid can be refined by using more elements of the same type but of smaller size (*h-refinement;* Cook et al., 1989) or by using elements of different type (*p-refinement*).

Formulation error results from use of finite elements that do not precisely describe the behavior of the continuum. For instance, a particular element might be formulated on the assumption that displacements vary linearly over the domain. Such an element would contain no formulation error when used to model a prismatic bar under constant tensile load; in this case, the assumed displacement matches the actual displacement. If the same bar were subjected to uniformly distributed body force, the actual displacements vary quadratically and formulation error would exist. Formulation error can be minimized by proper se lection of element type and by appropriate grid refinement.

Numerical error is a consequence of round-off during floating-point computations and the error associated with numerical integration procedures. This source of error is dependent on the order in which computations are performed in the program and on the use of double or extended-precision variables and functions. The use of bandwidth minimization[2] can help control numerical error. Generally, in a well-designed finite element program, numerical error is small relative to formulation error.

---

[2] See Section 4.6 for a discussion of bandwidth minimization.

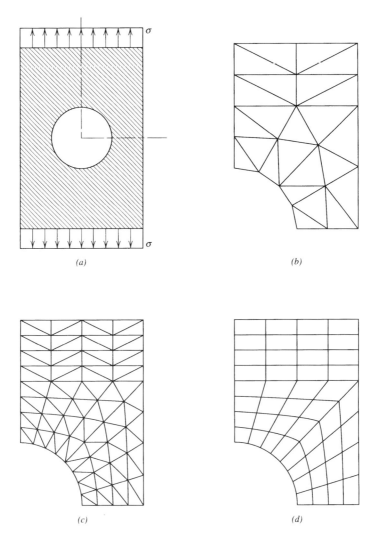

Figure 4.1   Finite element models of plate with centrally located hole. (a) Plate geometry and loading. (b) Coarse mesh of triangles. (c) Fine mesh of triangles. (d) Mesh of quadrilaterals with curved edges.

4.2   FORMULATION FOR PLANE ELASTICITY

Elasticity Concepts

One approach for developing the algebraic equations of the finite element method is to use energy principles. Fundamental energy expressions for an elastic solid can be found in standard texts (e.g., Boresi and Chong, 2000). For plane elasticity, these expressions are simplified appropriately. The first law of thermodynamics states that for a two-dimensional body in equilibrium and subjected to arbitrary virtual displacements (δu, δv), the variation in work of the external forces δW is equal to the variation of internal energy δU. Since *virtual displacements* are imposed, we define δW as the *virtual work of the external loads* and δU as the *virtual work of the internal forces*. The virtual work of the external forces δW can be divided into the work δW_S of the surface tractions, the work δW_B of the body forces, and the work $\delta W_C{}^3$ of the concentrated forces. For a two-dimensional body, these quantities are (Boresi et al., 1993, Chapters 3 and 5)

$$\delta W - \delta U = \delta W_S + \delta W_B + \delta W_C - \delta U = 0 \tag{4.1}$$

$$\delta W_S = \int_S (\sigma_{Px}\,\delta u + \sigma_{Py}\,\delta v)\,dS \tag{4.2}$$

$$\delta W_B = \int_V (B_x\,\delta u + B_y\,\delta v)\,dV \tag{4.3}$$

$$\delta W_C = \sum F_{ix}\,\delta u_i + \sum F_{iy}\,\delta v_i \tag{4.4}$$

$$\delta U = \int_V (\sigma_{xx}\,\delta\epsilon_{xx} + \sigma_{yy}\,\delta\epsilon_{yy} + \sigma_{xy}\,\delta\gamma_{xy})\,dV \tag{4.5}$$

where σ_{Px} and σ_{Py} are projections of the stress vector along axes x and y, respectively; B_x and B_y are body force components per unit volume; F_{ix} and F_{iy} are the x and y components of the concentrated force F_i at point i; δu_i and δv_i are the x and y components of the virtual displacement at point i; and $\gamma_{xy} = 2\epsilon_{xy}$. In matrix notation,

$$\delta W_S = \int_S \{\delta u\}^{\mathrm{T}}\{F_S\}\,dS \tag{4.6}$$

$$\delta W_B = \int_V \{\delta u\}^{\mathrm{T}}\{F_B\}\,dV \tag{4.7}$$

$$\delta W_C = \sum \{\delta u_i\}^{\mathrm{T}}\{F_i\} \tag{4.8}$$

$$\delta U = \int_V \{\delta\epsilon\}^{\mathrm{T}}\{\sigma\}\,dV \tag{4.9}$$

where

---

[3] Concentrated forces were not discussed in Chapter 3 or 5 of Boresi et al., 1993, but are included here for completeness.

$$\{\delta u\} = [\delta u \quad \delta v]^{\mathrm{T}}$$

$$\{\delta u_i\} = [\delta u_i \quad \delta v_i]^{\mathrm{T}}$$

$$\{F_S\} = [\sigma_{Px} \quad \sigma_{Py}]^{\mathrm{T}}$$

$$\{F_B\} = [B_x \quad B_y]^{\mathrm{T}}$$

$$\{F_i\} = [F_{ix} \quad F_{iy}]^{\mathrm{T}}$$

$$\{\delta\epsilon\} = [\delta\epsilon_{xx} \quad \delta\epsilon_{yy} \quad \delta\gamma_{xy}]^{\mathrm{T}}$$

$$\{\sigma\} = [\sigma_{xx} \quad \sigma_{yy} \quad \sigma_{xy}]^{\mathrm{T}}$$

In matrix form, the two-dimensional stress–strain relations are, by appropriate simplification [cf. Eq. (3.32); Boresi et al., 1993],

$$\{\sigma\} = [D]\{\epsilon\} \tag{4.10}$$

where $\{\epsilon\} = [\epsilon_{xx} \quad \epsilon_{yy} \quad \gamma_{xy}]^{\mathrm{T}}$ and $[D]$ is the matrix of elastic coefficients. For plane stress,

$$[D] = \frac{E}{1 - \nu^2} \begin{bmatrix} 1 & \nu & 0 \\ \nu & 1 & 0 \\ 0 & 0 & 1 - \dfrac{\nu}{2} \end{bmatrix} \tag{4.11}$$

and for plane strain,

$$[D] = \frac{E}{(1 + \nu)(1 - 2\nu)} \begin{bmatrix} 1 - \nu & \nu & 0 \\ \nu & 1 - \nu & 0 \\ 0 & 0 & \dfrac{1 - 2\nu}{2} \end{bmatrix} \tag{4.12}$$

Similarly, the two-dimensional, small displacement, strain–displacement relations are [see Eq. (2.81); Boresi et al., 1993]

$$\{\epsilon\} = [L]\{u\} \tag{4.13}$$

where $\{u\} = [u(x, y) \quad v(x, y)]^{\mathrm{T}}$ and $[L]$ is a matrix of linear differential operators:

$$[L] = \begin{bmatrix} \dfrac{\partial}{\partial x} & 0 \\[2mm] 0 & \dfrac{\partial}{\partial y} \\[2mm] \dfrac{\partial}{\partial y} & \dfrac{\partial}{\partial x} \end{bmatrix} \tag{4.14}$$

Displacement Interpolation: Constant-Strain Triangle

Consider a plane elasticity problem such as that shown in Fig. 4.1a. As discussed above, the first step in applying the finite element method is the discretization of the domain into a finite number of elements. Consider triangular elements as shown in Fig. 4.1b and c. If the entire domain is in equilibrium, so too is each of the elements. Hence, the virtual work concepts above can be applied to an individual triangular element.

A typical triangular element is shown in Fig. 4.2 with corner nodes 1, 2, and 3 numbered in a counterclockwise order. The (x, y) displacement components at the nodes are (u_1, v_1), (u_2, v_2), and (u_3, v_3), as shown. The *nodal* displacements are the primary variables (unknowns) that are to be determined by the finite element method. In general, for plane elasticity elements, node i has two degrees of freedom (DOFs), u_i and v_i, where the subscript identifies the node at which the DOF exists. Quantities that are continuous over the element (those not as-

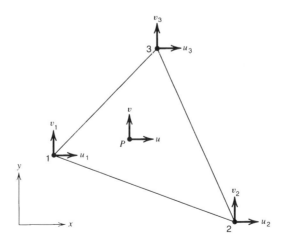

Figure 4.2 Constant-strain triangle element.

sociated with a particular node) are denoted without a subscript. A single triangular element with three nodes has six nodal DOFs These DOFs are ordered according to the node numbering as

$$\{u_i\} = [u_1 \quad v_1 \quad u_2 \quad v_2 \quad u_3 \quad v_3]^T \qquad (4.15)$$

Displacements (u, v) at any point P within the element are continuous functions of the spatial coordinates (x, y).

A fundamental approximation in the finite element method (that leads to formulation error) is that the displacement (u, v) at any point P in the element can be written in terms of the nodal displacements. Specifically, the displacement (u, v) at point P within the element is *interpolated* from the displacements of the nodes using interpolation polynomials. The order of the interpolation depends on the number of DOFs in the element. For the three-node triangular element, the displacement is assumed to vary linearly over the element.

$$u(x, y) = a_1 + a_2 x + a_3 y$$
$$v(x, y) = a_4 + a_5 x + a_6 y \qquad (4.16)$$

The coefficients a_i are constants (sometimes called *generalized displacement coordinates*) that are evaluated in terms of the nodal displacements. Before making this evaluation, we consider some properties of the linear displacement approximation.

1. Substitution of Eq. (4.16) into Eq. (4.13) yields $\epsilon_{xx} = a_2$, $\epsilon_{yy} = a_6$, and $\gamma_{xy} = a_3 + a_5$. Thus, the strain components in the element are constant: hence the name *constant-strain triangle* (CST) element. Since the stress–strain relations are linear [Eq. (4.10)], stress components are also constant in the element.

2. If $a_2 = a_3 = a_5 = a_6 = 0$, then $u(x, y) = a_1$ and $v(x, y) = a_4$. Constant values of u and v displacement indicate that the element can represent rigid-body translation.

3. If $a_1 = a_2 = a_4 = a_6 = 0$ and $a_3 = -a_5$, then $u(x, y) = a_3 y$ and $v(x, y) = -a_3 x$. Thus, the element can represent rigid-body rotation.

These three element characteristics ensure that the solution will converge monotonically as the mesh is refined (see Section 4.6 for a discussion of convergence).

To express the continuous displacement field in terms of the nodal displacements, Eq. (4.16) is evaluated at each node. The resulting equations are then solved for the coefficients a_i. Consider first the u displacement.

$$u(x_1, y_1) = a_1 + a_2 x_1 + a_3 y_1 = u_1$$

$$u(x_2, y_2) = a_1 + a_2 x_2 + a_3 y_2 = u_2$$

$$u(x_3, y_3) = a_1 + a_2 x_3 + a_3 y_3 = u_3$$

In matrix form, these equations are written as

$$[A]\{a\} = \{u_i\} \tag{4.17}$$

where

$$[A] = \begin{bmatrix} 1 & x_1 & y_1 \\ 1 & x_2 & y_2 \\ 1 & x_3 & y_3 \end{bmatrix} \qquad \{a\} = \begin{Bmatrix} a_1 \\ a_2 \\ a_3 \end{Bmatrix} \qquad \{u_i\} = \begin{Bmatrix} u_1 \\ u_2 \\ u_3 \end{Bmatrix}$$

Solution of Eq. (4.17) for $\{a\}$ and substitution into Eq. (4.16) yields

$$u(x, y) = \tfrac{1}{2A}(\alpha_1 + \beta_1 x + \gamma_1 y)u_1 + \tfrac{1}{2A}(\alpha_2 + \beta_2 x + \gamma_2 y)u_2$$
$$+ \tfrac{1}{2A}(\alpha_3 + \beta_3 x + \gamma_3 y)u_3 \tag{4.18}$$

where A is the area of the triangle:

$$A = \tfrac{1}{2}[x_1(y_2 - y_3) + x_2(y_3 - y_1) + x_3(y_1 - y_2)] \tag{4.19}$$

and

$$\begin{array}{lll} \alpha_1 = x_2 y_3 - x_3 y_2 & \beta_1 = y_2 - y_3 & \gamma_1 = x_3 - x_2 \\ \alpha_2 = x_3 y_1 - x_1 y_3 & \beta_2 = y_3 - y_1 & \gamma_2 = x_1 - x_3 \quad (4.20) \\ \alpha_3 = x_1 y_2 - x_2 y_1 & \beta_3 = y_1 - y_2 & \gamma_3 = x_2 - x_1 \end{array}$$

Similarly, for the v displacement,

$$v(x, y) = \tfrac{1}{2A}(\alpha_1 + \beta_1 x + \gamma_1 y)v_1 + \tfrac{1}{2A}(\alpha_2 + \beta_2 x + \gamma_2 y)v_2$$

$$+ \tfrac{1}{2A}(\alpha_3 + \beta_3 x + \gamma_3 y)v_3 \tag{4.21}$$

The functions that multiply the nodal displacements in Eqs. (4.18) and (4.21) are known as *shape functions* (other common names are *interpolation functions* and *basis functions*). The shape functions for the CST element are

$$N_1(x, x) = \tfrac{1}{2A}(\alpha_1 + \beta_1 x + \gamma_1 y)$$

$$N_2(x, y) = \tfrac{1}{2A}(\alpha_2 + \beta_2 x + \gamma_2 y) \tag{4.22}$$

$$N_3(x, y) = \tfrac{1}{2A}(\alpha_3 + \beta_3 x + \gamma_3 y)$$

Then Eqs. (4.18) and (4.21) take the form

$$u(x, y) = \sum_{i=1}^{3} N_i u_i \qquad v(x, y) = \sum_{i=1}^{3} N_i v_i$$

In matrix notation

$$\{u\} = [N]\{u_i\} \tag{4.23}$$

where

$$[N] = \begin{bmatrix} N_1 & 0 & N_2 & 0 & N_3 & 0 \\ 0 & N_1 & 0 & N_2 & 0 & N_3 \end{bmatrix} \tag{4.24}$$

The shape functions are illustrated in Fig. 4.3, where $N_1 = 1$ at

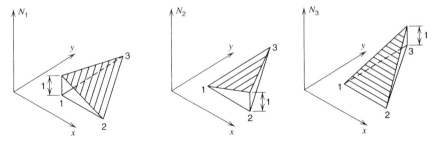

Figure 4.3 Graphical representation of shape functions for the CST element.

node 1 and $N_1 = 0$ at nodes 2 and 3. Shape functions N_2 and N_3 behave similarly. Another important characteristic of the shape functions is that

$$\sum_{i=1}^{3} N_i(x, y) = 1.0$$

which is a consequence of the fact that the element can represent rigid-body motion.

Element Stiffness Matrix: Constant-Strain Triangle

With the displacement field for the element expressed in terms of the nodal displacements, the remainder of the formulation involves relatively straightforward manipulation of the virtual work expressions, Eqs. (4.1), (4.6)–(4.9). Consider first the strain–displacement relations. Substitution of Eq. (4.23) into Eq. (4.13) gives the relationship between *continuous element strains* and *nodal displacements,*

$$\{\epsilon\} = [L][N]\{u_i\} = [B]\{u_i\} \tag{4.25}$$

where, by Eqs. (4.14) and (4.24),

$$[B] = \begin{bmatrix} \dfrac{\partial N_1}{\partial x} & 0 & \bigg| & \dfrac{\partial N_2}{\partial x} & 0 & \bigg| & \dfrac{\partial N_3}{\partial x} & 0 \\[2mm] 0 & \dfrac{\partial N_1}{\partial y} & \bigg| & 0 & \dfrac{\partial N_2}{\partial y} & \bigg| & 0 & \dfrac{\partial N_3}{\partial y} \\[2mm] \dfrac{\partial N_1}{\partial y} & \dfrac{\partial N_1}{\partial x} & \bigg| & \dfrac{\partial N_2}{\partial y} & \dfrac{\partial N_2}{\partial x} & \bigg| & \dfrac{\partial N_3}{\partial y} & \dfrac{\partial N_3}{\partial x} \end{bmatrix} \tag{4.26}$$

where $[B]$ is partitioned into nodal submatrices. The matrix $[B]$ is sometimes called the *semidiscretized gradient operator.* Since the shape functions are linear in x and y, $[B]$ contains only constants that depend on the nodal coordinates.

  For simplicity, temporarily assume that no body forces or surface tractions are applied to the element. However, concentrated loads at node points are permitted. The virtual work of these loads is [see Eq. (4.8)]

$$\delta W = \delta W_C = \{\delta u_i\}^T\{F_i\} \tag{4.27}$$

Substitution of Eqs. (4.27) and (4.9) into (4.1) leads to

$$\{\delta u_i\}^T\{F_i\} - \int_V \{\delta\epsilon\}^T\{\sigma\}\,dV = 0 \tag{4.28}$$

Note that $\{\sigma\} = [D]\{\epsilon\}$, $\{\epsilon\} = [B]\{u_i\}$, and $\{\delta\epsilon\}^T = \{\delta u_i\}^T[B]^T$. Substitution of these expressions into Eq. (4.28) gives

$$\{\delta u_i\}^T\{F_i\} - \int_V \{\delta u_i\}^T[B]^T[D][B]\{u_i\}\,dV = 0$$

Since $\{u_i\}$ and $\{\delta u_i\}$ are *nodal* quantities, they can be removed from the integral. Thus

$$\{\delta u_i\}^T\left(\{F_i\} - \left[\int_V [B]^T[D][B]\,dV\right]\{u_i\}\right) = 0 \tag{4.29}$$

Since $\{\delta u_i\}$ is arbitrary, Eq. (4.29) becomes

$$\{F_i\} = \left[\int_V [B]^T[D][B]\,dV\right]\{u_i\}$$

or

$$\{F_i\} = [K]\{u_i\} \tag{4.30}$$

where

$$[K] = \int_V [B]^T[D][B]\,dV \tag{4.31}$$

The element stiffness matrix $[K]$ relates nodal loads to nodal displacements in a system of linear algebraic equations, Eq. (4.30). For the CST element, all terms in the integral are constants. Hence, for an element of constant thickness t, the element stiffness matrix is

$$[K] = At[B]^T[D][B] \tag{4.32}$$

The individual terms in $[K]$ are denoted k_{ij}, where $i, j = 1, 2, \ldots, 6$

are the row and column positions, respectively. Since the element has six nodal DOF, $[K]$ has order (6×6). The explicit form of the CST element stiffness matrix for a plane stress condition is given in Table 4.1.

Examination of Eq. (4.30) helps to establish a physical interpretation of the stiffness coefficients (the individual terms in $[K]$). Let a unit displacement be assigned to u_1 and take all other DOFs to be zero. The resulting displacement vector is

$$\{u_i\} = [1 \quad 0 \quad 0 \quad 0 \quad 0 \quad 0]^T$$

Substitution of this displacement vector into Eq. (4.30) gives the force vector required to maintain the deformed shape:

$$\{F_i\} = [k_{11} \quad k_{21} \quad k_{31} \quad k_{41} \quad k_{51} \quad k_{61}]^T$$

Hence, an individual stiffness coefficient k_{ij} can be interpreted as the nodal force in the direction of DOF i that results from a unit displacement in the direction of DOF j while all other DOF are set equal to zero. The physical system is illustrated in Fig. 4.4.

Equivalent Nodal Load Vector: Constant-Strain Triangle

Assume that body forces are applied to the CST element (surface tractions will be considered subsequently). The virtual work δW_B of the body forces on the element during an arbitrary virtual displacement $\{\delta u\}$ is given by Eq. (4.7). Substitution of Eq. (4.23) into Eq. (4.7) gives

$$\delta W_B = \{\delta u_i\}^T \int_V [N]^T \{F_B\} \, dV \tag{4.33}$$

The total external virtual work δW is the sum of the virtual work of the body forces and the virtual work of the concentrated forces, so that Eq. (4.29) becomes

$$\{\delta u_i\}^T \left(\{F_i\} + \int_V [N]^T \{F_B\} \, dV - \left[\int_V [B]^T [D][B] \, dV \right] \{u_i\} \right) = 0$$

$$\tag{4.34}$$

TABLE 4.1 CST Element Stiffness Matrix, Plane Stress Case (Partitioned into 2 × 2 Nodal Submatrices)

column index

$j \rightarrow$

1	2	3	4	5	6	row index $i\downarrow$
$y_{23}^2 + \dfrac{1-\nu}{2} x_{32}^2$	$\dfrac{1+\nu}{2} x_{32} y_{23}$	$y_{31} y_{23} + \dfrac{1+\nu}{2} x_{13} x_{32}$	$\nu x_{13} y_{23} + \dfrac{1-\nu}{2} x_{32} y_{31}$	$y_{12} y_{23} + \dfrac{1-\nu}{2} x_{21} x_{32}$	$\nu x_{21} y_{23} + \dfrac{1-\nu}{2} x_{32} y_{12}$	1
$\dfrac{1+\nu}{2} x_{32} y_{23}$	$x_{32}^2 + \dfrac{1-\nu}{2} y_{23}^2$	$\nu x_{32} y_{31} + \dfrac{1-\nu}{2} x_{13} y_{23}$	$x_{13} x_{32} + \dfrac{1-\nu}{2} y_{23} y_{31}$	$\nu x_{32} y_{12} + \dfrac{1-\nu}{2} x_{21} y_{23}$	$x_{21} x_{32} + \dfrac{1-\nu}{2} y_{12} y_{23}$	2
$y_{31} y_{23} + \dfrac{1+\nu}{2} x_{13} x_{32}$	$\nu x_{32} y_{31} + \dfrac{1-\nu}{2} x_{13} y_{23}$	$y_{31}^2 + \dfrac{1-\nu}{2} x_{13}^2$	$\dfrac{1+\nu}{2} x_{13} y_{31}$	$y_{12} y_{31} + \dfrac{1-\nu}{2} x_{13} x_{21}$	$\nu x_{21} y_{31} + \dfrac{1-\nu}{2} x_{13} y_{12}$	3
$\nu x_{13} y_{23} + \dfrac{1-\nu}{2} x_{32} y_{31}$	$x_{13} x_{32} + \dfrac{1-\nu}{2} y_{23} y_{31}$	$\dfrac{1+\nu}{2} x_{13} y_{31}$	$x_{13}^2 + \dfrac{1-\nu}{2} y_{31}^2$	$\nu x_{13} y_{12} + \dfrac{1-\nu}{2} x_{21} y_{31}$	$x_{13} x_{21} + \dfrac{1-\nu}{2} y_{12} y_{31}$	4
$y_{12} y_{23} + \dfrac{1-\nu}{2} x_{21} x_{32}$	$\nu x_{32} y_{12} + \dfrac{1-\nu}{2} x_{21} y_{23}$	$y_{12} y_{31} + \dfrac{1-\nu}{2} x_{13} x_{21}$	$\nu x_{13} y_{12} + \dfrac{1-\nu}{2} x_{21} y_{31}$	$y_{12}^2 + \dfrac{1-\nu}{2} x_{21}^2$	$\dfrac{1+\nu}{2} x_{21} y_{12}$	5
$\nu x_{21} y_{23} + \dfrac{1-\nu}{2} x_{32} y_{12}$	$x_{21} x_{32} + \dfrac{1-\nu}{2} y_{12} y_{23}$	$\nu x_{21} y_{31} + \dfrac{1-\nu}{2} x_{13} y_{12}$	$x_{13} x_{21} + \dfrac{1-\nu}{2} y_{12} y_{31}$	$\dfrac{1+\nu}{2} x_{21} y_{12}$	$x_{21}^2 + \dfrac{1-\nu}{2} y_{12}^2$	6

C

$$C = \frac{Et}{4A(1-\nu^2)} \qquad x_{ij} = x_i - x_j \qquad y_{ij} = y_i - y_j$$

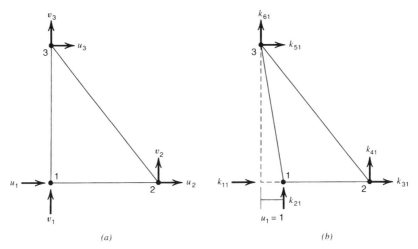

Figure 4.4 Physical interpretation of k_{ij}. (a) Undeformed element. (b) Deformed element, forces k_{i1} required to maintain $u_1 = 1$.

Comparison of Eq. (4.34) with Eq. (4.29) shows that with the addition of body forces, the load vector for the element becomes

$$\{P_i\} = \{F_i\} + \int_V [N]^T \{F_B\} \, dV \tag{4.35}$$

The vector $\{P_i\}$ is the equivalent nodal load vector for the element. That is, the work of the loads $\{P_i\}$ under the virtual displacement $\{\delta u_i\}$ of the nodes is equivalent to the work of the actual concentrated loads and body forces under the virtual displacement $\{\delta u\}$.

In Eq. (4.35), the body force $\{F_B\}$ is expressed as a continuous function of the spatial coordinates. However, when constructing a finite element model, it is customary for the analyst to define element loads in terms of the intensity of the load at the nodes rather than in functional form. So, for convenience, assume that the body force distribution may be expressed in terms of the force intensities at the nodes according to the relation

$$\{F_B\} = [N]\{f_{Bi}\}$$

where $\{f_{Bi}\}$ is the vector of nodal force intensities. Substitution of this relation into Eq. (4.35) gives

$$\{P_i\} = \{F_i\} + \int_V [N]^T [N]\{f_{Bi}\} \, dV$$

Since $\{f_{Bi}\}$ does not vary over the element,

$$\{P_i\} = \{F_i\} + [Q]\{f_{Bi}\} \tag{4.36}$$

where

$$[Q] = \int_V [N]^T[N] \, dV$$

Thus, for a CST element,

$$[Q] = \frac{At}{9} \begin{bmatrix} 1 & 0 & 1 & 0 & 1 & 0 \\ 0 & 1 & 0 & 1 & 0 & 1 \\ 1 & 0 & 1 & 0 & 1 & 0 \\ 0 & 1 & 0 & 1 & 0 & 1 \\ 1 & 0 & 1 & 0 & 1 & 0 \\ 0 & 1 & 0 & 1 & 0 & 1 \end{bmatrix}$$

Now suppose that in addition to concentrated nodal loads and body forces, the element is subjected to surface tractions along a single edge and that the continuous load function $\{F_S\}$ is expressed in terms of the nodal force intensities $\{f_{Si}\}$ by use of the shape functions. Since only one edge is loaded, only two of the nodes have nodal intensities and only these two nodes have equivalent nodal load components. Hence, for these two nodes, the interpolation equation is

$$\{\bar{F}_S\} = [\bar{N}]\{\bar{f}_{Si}\}$$

where the overbar indicates that only these two element nodes are included in the equation.

By the same approach as for body forces, the equivalent nodal loads due to surface traction on one edge are

$$\{\bar{P}_i\} = [\bar{Q}]\{\bar{f}_{Si}\} \tag{4.37}$$

where

$$[\bar{Q}] = \int_S [\bar{N}]^T[\bar{N}] \, dS \tag{4.38}$$

and the integral is evaluated over the loaded edge only, where $dS = t \, ds$, $t = $ thickness, and s is a coordinate along the loaded edge. The equivalent nodal load vector $\{\overline{P}_i\}$ in Eq. (4.37) is then added to $\{P_i\}$ from Eq. (4.36), but first it must be expanded from four to six terms to account for the fact that one node does not participate in the loading.

Example 4.1:Equivalent Nodal Loads for Linear Surface Traction
A horizontally directed, linearly varying surface traction is applied to edge 1–3 of the CST element with nodal intensities as shown in Fig. E4.1. Determine the vector of equivalent nodal loads for the element.

SOLUTION   The surface traction function is interpolated from the nodal intensities at nodes 1 and 3 and from the corresponding shape functions

$$f_x(y) = N_1 f_{1x} + N_2 f_{3x} \tag{a}$$

With the coordinates of the nodes, the shape functions are simplified to

$$N_1 = 1 - \frac{y}{b} \qquad N_3 = \frac{y}{b} \tag{b}$$

By Eq. (4.38), with $ds = dy$,

$$[\overline{Q}] = t \int_0^b \begin{bmatrix} N_1^2 & 0 & N_1 N_3 & 0 \\ 0 & N_1^2 & 0 & N_1 N_3 \\ N_1 N_3 & 0 & N_3^2 & 0 \\ 0 & N_1 N_3 & 0 & N_3^2 \end{bmatrix} dy \tag{c}$$

By Eqs. (b) and (c),

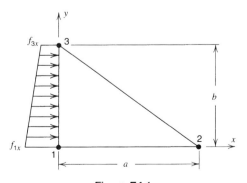

Figure E4.1

$$[\overline{Q}] = t \begin{bmatrix} b/3 & 0 & b/6 & 0 \\ 0 & b/3 & 0 & b/6 \\ b/6 & 0 & b/3 & 0 \\ 0 & b/6 & 0 & b/3 \end{bmatrix} \tag{d}$$

and by Eq. (a), the vector of nodal intensities $\{\overline{f}_{Si}\}$ is

$$\{\overline{f}_{Si}\} = [f_{1x} \quad 0 \quad f_{3x} \quad 0]^{\mathrm{T}} \tag{e}$$

With Eqs. (d) and (e), the equivalent nodal load vector $\{\overline{P}_i\}$ is obtained from Eq. (4.37) as

$$\{\overline{P}_i\} = \left\{ \begin{array}{c} tb\left(\dfrac{f_{1x}}{3} + \dfrac{f_{3x}}{6}\right) \\[2mm] 0 \\[2mm] tb\left(\dfrac{f_{1x}}{6} + \dfrac{f_{3x}}{3}\right) \\[2mm] 0 \end{array} \right\} \tag{f}$$

Equation (e) is partitioned to identify the equivalent nodal loads associated with nodes 1 and 3.

If the vector $\{\overline{P}_i\}$ is expanded to include positions for node 2, it becomes

$$\{\overline{P}_i\} = \left\{ \begin{array}{c} tb\left(\dfrac{f_{1x}}{3} + \dfrac{f_{3x}}{6}\right) \\[2mm] 0 \\ \hline 0 \\ 0 \\ \hline tb\left(\dfrac{f_{1x}}{6} + \dfrac{f_{3x}}{3}\right) \\[2mm] 0 \end{array} \right\} \tag{g}$$

Assembly of the Structure Stiffness Matrix and Load Vector

To solve a plane elasticity problem by the finite element method, it is necessary to combine the individual element stiffness matrices $[K]_j$ and load vectors $\{P\}_j$ to form the *structure stiffness matrix* [**K**] and the *structure load vector* {**P**}, respectively. To demonstrate the logic associated with the assembly process, two node numbering systems for the nodes are used. Let numerals in boldface refer to the nodes of the structural system and numerals in lightface refer to the nodes for a particular element. Similarly, lightface [K], $\{u\}$, and $\{P\}$ refer to element quantities and boldface [**K**], {**u**}, and {**P**} refer to structure quantities. A specific two-dimensional discretization is shown in Fig. 4.5 to illustrate the node numbering. For this model, there are six structure nodes but a total of twelve separate element nodes. The assembly process involves assigning unique identifiers to each of the nodes in the model, using the structure node numbering and then combining element stiffness matrices and load vectors according to the numbering.

For purposes of demonstration, we consider first a mathematically precise but computationally inefficient approach for this assembly. Then we discuss an approach that is more appropriate for computer implementation.

For element j, define a matrix $[M]_j$ with order $(6 \times 2NN)$,[4] where NN is the number of nodes in the structure, to define the mapping from the element DOF vector $\{u\}_j$, with order (6×1), to the structure DOF vector {**u**}, with order $(2NN \times 1)$:

$$\{u\}_j = [M]_j\{\mathbf{u}_i\} \tag{4.39}$$

By Fig. 4.5, the mapping for element 1 takes the form

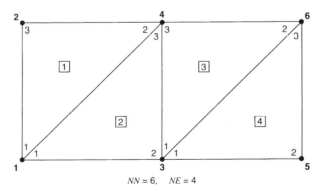

$$NN = 6, \quad NE = 4$$

Figure 4.5 Assembly of CST elements.

---

[4]The matrix $[M]_j$ is known as a *Boolean connectivity matrix* since it contains only ones and zeros.

$$\{u\}_1 = [u_1 \quad v_1 \quad u_2 \quad v_2 \quad u_3 \quad v_3]_1^T$$

$$\{\mathbf{u}\} = [u_1 \quad v_1 \quad u_2 \quad v_2 \quad u_3 \quad v_3 \quad \ldots \quad v_6]^T$$

and

$$[M]_1 = \begin{bmatrix} 1 & 0 & 0 & 0 & 0 & 0 & 0 & 0 & 0 & 0 & 0 & 0 \\ 0 & 1 & 0 & 0 & 0 & 0 & 0 & 0 & 0 & 0 & 0 & 0 \\ 0 & 0 & 0 & 0 & 0 & 0 & 1 & 0 & 0 & 0 & 0 & 0 \\ 0 & 0 & 0 & 0 & 0 & 0 & 0 & 1 & 0 & 0 & 0 & 0 \\ 0 & 0 & 1 & 0 & 0 & 0 & 0 & 0 & 0 & 0 & 0 & 0 \\ 0 & 0 & 0 & 1 & 0 & 0 & 0 & 0 & 0 & 0 & 0 & 0 \end{bmatrix}$$

By inspection or by Eq. (4.39), the DOF mapping for element 1 is

$$[u_1 \quad v_1 \quad u_2 \quad v_2 \quad u_3 \quad v_3]_1^T \leftrightarrow [u_1 \quad v_1 \quad u_4 \quad v_4 \quad u_2 \quad v_2]^T$$

The double-headed arrow indicates the reversibility of the mapping of the quantities on the left to the quantities on the right. Nodal forces and stiffness coefficients for element 1 follow the same mapping.

Next, the virtual work expressions for the entire structure are written as the sum of the virtual work for all elements.

$$\sum_{j=1}^{NE} \{\delta u_i\}_j^T \{F_i\}_j + \sum_{j=1}^{NE} \{\delta u_i\}_j^T \int_S [\overline{N}]_j^T \{F_S\}_j \, dS$$

$$+ \sum_{j=1}^{NE} \{\delta u_i\}_j^T \int_V [N]_j^T \{F_B\}_j \, dV - \sum_{j=1}^{NE} \{\delta u_i\}_j^T [K]_j \{u_i\}_j = 0 \quad (4.40)$$

where NE is the number of elements in the model. Substitution of Eq. (4.39) into Eq. (4.40) for each element gives

$$\{\delta \mathbf{u}_i\}^T \sum_{j=1}^{NE} [M]_j^T \{F_i\}_j + \{\delta \mathbf{u}_i\}^T \sum_{j=1}^{NE} [\overline{M}]_j^T \int_S [\overline{N}]_j^T \{F_S\}_j \, dS$$

$$+ \{\delta \mathbf{u}_i\}^T \sum_{j=1}^{NE} [M]_j^T \int_V [N]_j^T \{F_B\}_j \, dV$$

$$- \{\delta \mathbf{u}_i\}^T \left[\sum_{j=1}^{NE} [M]_j^T [K]_j [M]_j \right] \{\mathbf{u}_i\} \quad (4.41)$$

$$= 0$$

Since $\{\delta \mathbf{u}_i\}$ is arbitrary, it is eliminated from Eq. (4.41) to obtain

$$[\mathbf{K}]\{\mathbf{u}_i\} = \{\mathbf{P}_i\} \tag{4.42a}$$

where

$$[\mathbf{K}] = \left[\sum_{j=1}^{NE} [M]_j^T [K]_j [M]_j \right] \tag{4.42b}$$

and

$$\{\mathbf{P}_i\} = \sum_{j=1}^{NE} [M]_j^T \{F_i\}_j + \sum_{j=1}^{NE} [\overline{M}]_j^T \int_S [\overline{N}]_j^T \{F_S\}_j \, dS$$

$$+ \sum_{j=1}^{NE} [M]_j^T \int_V [N]_j^T \{F_B\}_j \, dV \tag{4.42c}$$

In Eqs. (4.41) and (4.42c), matrix $[\overline{M}]_j$, of order $(4 \times 2NN)$, accounts for the mapping to the structure nodes of the two nodes in element j that participate in the surface tractions. If more than one edge on an element is loaded, Eqs. (4.41) and (4.42b) are extended accordingly.

The forms of $[\mathbf{K}]$ and $\{\mathbf{P}_i\}$ in Eq. (4.42) are precise, but they are not used in practice. The matrix products involving $[M]_j$, which involve multiplying by 0 or 1, do nothing more than move individual quantities from one position in the element stiffness matrix or load vector to another in the structure stiffness matrix or load vector. Although the development noted above is not practical, it does demonstrate that the structure stiffness matrix is assembled by successively adding the stiffness terms from each element into appropriate locations of the structure matrix, and similarly for the structure load vector. The more direct approach to assembly is demonstrated in Example 4.2.

Example 4.2: Assembly of the Structure Stiffness Matrix   For the model shown in Fig. 4.5, illustrate the assembly of the stiffness matrix for element 1 into the structure stiffness matrix.

SOLUTION   Since the structure has six nodes, each of which has two DOFs, the structure stiffness matrix is of order 12×12. The individual stiffness coefficients are designated k_{ij}^e, where the superscript identifies the element number. With this notation the stiffness matrix for element 1 is shown in Fig. E4.2. The mapping of element node numbers to structure node numbers is determined by inspection of the model in Fig. 4.5 and is summarized in Table E4.2. The list of structure node

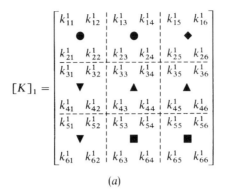

$$(a)$$

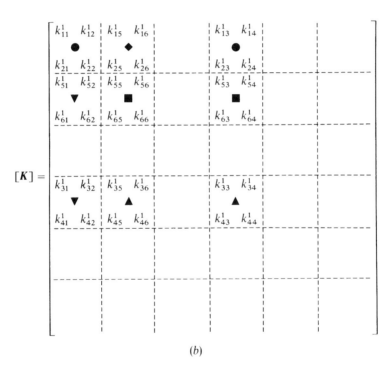

$$(b)$$

Figure E4.2 Assembly of element 1 stiffness matrix (with 2 × 2 nodal submatrix partitions). (a) Stiffness matrix for element 1. (b) Structure stiffness matrix with element 1 assembled.

numbers that define the nodes for each element, commonly known as the incidence list, is one of the input requirements for finite element programs. Using the incidence list, one can obtain the mapping of the element 1 nodal submatrices into the structure stiffness matrix (Fig. E4.2). Markers have been added to the nodal submatrices as an aid to

TABLE E4.2 Element Node-to-Structure Node Mapping

Element Node Number	Structure Node Numbers			
	Element 1	Element 2	Element 3	Element 4
1	1	1	3	3
2	4	3	6	5
3	2	4	4	6

visualization of the placement of element stiffness coefficients into the structure stiffness matrix.

As described in Example 4.2, the incidence list is used to drive the assembly process. Suppose that the node numbers that comprise the incidence list, for instance, Table E4.2, are placed into a matrix [INCID] that contains one column for each element. The i,j term in the matrix is defined as the structure node number that corresponds to element node number i of element j. Then, using the incidence matrix [INCID], each term from the element stiffness matrix $[K]_j$ is moved into the structure stiffness matrix $[\mathbf{K}]$ in a prescribed manner. The method is illustrated in Table 4.2 by a FORTRAN subroutine. The subroutine moves one nodal submatrix at a time. Note that this code is for illustrative purposes only. Because of the symmetry and sparsity of the structure stiffness, it is usually stored in some form other than a square matrix.

Application of Constraints

The model shown in Fig. 4.5 is not fastened to supports. Hence, it represents an unstable structure, a structure that is not capable of resisting external loads. The assembled stiffness matrix for an unstable structure is singular; it has a rank deficiency of 3, due to the three rigid-body modes that the model possesses. Physically, the structure must be supported to prevent rigid-body motion. In a similar fashion, if the structure stiffness matrix is modified to reflect the support conditions (commonly known as *constraints*), it becomes nonsingular. Several methods may be used to apply constraints to the structure stiffness matrix. Only one, the *equation modification method,* is discussed here.

To demonstrate the equation modification method, consider a model that contains only a single element (Fig. 4.6a). The first step is to switch appropriate rows and columns of the stiffness such that those DOFs that are constrained are grouped together. The rearranged stiff-

TABLE 4.2   FORTRAN Subroutine for Stiffness Assembly

```
      SUBROUTINE ASSMBL ( KS, KE, NNE, NDOF, INCID, IE, NRS, NRE, NE )
      DIMENSION KS ( NRS, NRS ), KE ( NRE, NR ), INCID ( NNE, NE )
      REAL KS, KE
C
C        ASSEMBLE THE STIFFNESS FOR ELEMENT 'IE' INTO THE
C        STRUCTURE STIFFNESS.
C
C     CONTROL VARIABLES:
C
C        KS, KE   = STRUCTURE & ELEMENT STIFFNESS MATRICES.
C        NNE      = NUMBER OF NODES IN AN ELEMENT.
C        NDOF     = NUMBER OF DOF AT EACH NODE.
C        INCID    = INCIDENCE MATRIX.
C        IE       = CURRENT ELEMENT NUMBER.
C        NRS, NRE = NUMBER OF ROWS IN STRUCTURE & ELEMENT STIFFNESS.
C        NE       = NUMBER OF ELEMENTS IN THE MODEL.
C
C     LOCAL VARIABLES:
C
C        INE      = CURRENT ELEMENT SUBMATRIX ROW NUMBER.
C        JNE      = CURRENT ELEMENT SUBMATRIX COLUMN NUMBER.
C        INS      = CURRENT STRUCTURE SUBMATRIX ROW NUMBER.
C        JNS      = CURRENT STRUCTURE SUBMATRIX COLUMN NUMBER.
C        IDOF     = CURRENT DOF NUMBER IN SUBMATRIX ROW.
C        JDOF     = CURRENT DOF NUMBER IN SUBMATRIX COLUMN.
C        IKE      = ROW ENTRY IN THE ELEMENT STIFFNESS.
C        JKE      = COLUMN ENTRY IN THE ELEMENT STIFFNESS.
C        IKS      = ROW ENTRY IN THE STRUCTURE STIFFNESS.
C        JKS      = COLUMN ENTRY IN THE STRUCTURE STIFFNESS.
C
      DO 10 INE = 1, NNE
         INS = INCID ( INE, IE )
      DO 10 JNE = 1, NNE
         JNS = INCID ( JNE, IE )
C
C        ASSEMBLE ELEMENT SUBMATRIX (INE, JNE) INTO STRUCTURE SUBMATRIX
C        (INS, JNS)
C
         DO 10 IDOF = 1, NDOF
            IKE = ( INE - 1 ) * NDOF + IDOF
            IKS = ( INS - 1 ) * NDOF + IDOF
         DO 10 JDOF - 1, NDOF
               JKE = ( JNE - 1 ) * NDOF + JDOF
               JKS = ( JNS - 1 ) * NDOF + JDOF
C
         KS( IKS, JKS ) = KS( IKS, JKS ) + KE( IKE, JKE )
C
   10       CONTINUE
            RETURN
            END
```

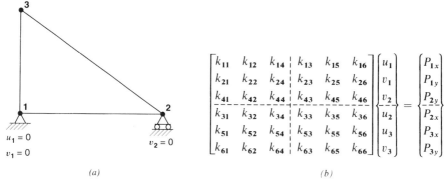

Figure 4.6 Application of constraints by the equation modification method. (a) One-element model with constraints. (b) Rearranged structure equations.

ness matrix, displacement vector, and load vector for the one-element model are shown in Fig. 4.6b. For simplicity, the rearranged equations are represented in the symbolic form

$$\left[\begin{array}{c|c} \mathbf{K}_{cc} & \mathbf{K}_{cu} \\ \hline \mathbf{K}_{uc} & \mathbf{K}_{uu} \end{array}\right] \left\{\begin{array}{c} \mathbf{u}_c \\ \mathbf{u}_u \end{array}\right\} = \left\{\begin{array}{c} \mathbf{P}_c \\ \mathbf{P}_u \end{array}\right\} \tag{4.43}$$

where the subscript C represents the constrained DOF and the subscript u represents the unconstrained DOF.[5] The relationship between the submatrices and subvectors in Eq. (4.43) and those in Fig. 4.6b is determined by their respective positions in the equations.

The unknown quantities are the displacements $\{\mathbf{u}_u\}$ of the unconstrained DOF and the forces $\{\mathbf{P}_c\}$ at the constrained DOF. Rewrite Eq. (4.43) as two separate submatrix/subvector equations.

$$[\mathbf{K}_{cc}]\{\mathbf{u}_c\} + [\mathbf{K}_{cu}]\{\mathbf{u}_u\} = \{\mathbf{P}_c\} \tag{4.44a}$$

$$[\mathbf{K}_{uc}]\{\mathbf{u}_c\} + [\mathbf{K}_{uu}]\{\mathbf{u}_u\} = \{\mathbf{P}_u\} \tag{4.44b}$$

Since $\{\mathbf{u}_c\}$ is known, it is moved to the load side of Eq. (19.44b), to obtain

---

[5] The rearrangement of the equations and subsequent partitioning is done for convenience in representing the method. Computer implementation of this approach does not require that the equations be rearranged. Nor would such rearrangement be computationally efficient.

$$[\mathbf{K}_{uu}]\{\mathbf{u}_u\} = \{\mathbf{P}_u\} - [\mathbf{K}_{uc}]\{\mathbf{u}_c\} \qquad (4.45)$$

Equation (4.45) is the constrained system of equations. If the imposed constraints $\{\mathbf{u}_c\}$ are nonzero, they serve to modify the load vector. If the constraints are all zero, such as in Fig. 4.6*a*, then the second term on the right side of Eq. (4.45) vanishes. In either case, the system of equations is reduced in order by the number of constrained DOFs. If appropriate constraints are applied to render the structure stable, $[\mathbf{K}_{uu}]$ will be nonsingular.

Solution of the System of Equations

After assembly of the stiffness matrix and load vector and application of constraints, the system of linear algebraic equations may be solved. It is common to represent the solution of Eq. (4.45) in the symbolic form

$$\{\mathbf{u}_u\} = [\mathbf{K}_{uu}]^{-1}(\{\mathbf{P}_u\} - [\mathbf{K}_{uc}]\{\mathbf{u}_c\})$$

However, inversion of the stiffness matrix $[\mathbf{K}_{uu}]$ is computationally expensive and can lead to significant numerical error. A more efficient approach, known as *Choleski decomposition,* involves triangular factorization of the stiffness matrix:

$$[\mathbf{K}_{uu}] = [\mathbf{U}]^{\mathrm{T}}[\mathbf{U}]$$

where $[\mathbf{U}]$ is an upper triangular matrix; that is, each term in the lower triangle of $[\mathbf{U}]$ is zero ($u_{ij} = 0$, $i > j$). Factorization of $[\mathbf{K}_{uu}]$ into this form permits direct solution for displacements via two *load-pass* operations. The first of these, known as the *forward load-pass,* yields an intermediate solution vector $\{\mathbf{y}\}$:

$$[\mathbf{U}]^{\mathrm{T}}\{\mathbf{y}\} = \{\mathbf{P}_u\} - [\mathbf{K}_{uc}]\{\mathbf{u}_c\}$$

The second operation, known as the *backward load pass,* yields the final displacement vector $\{\mathbf{u}_u\}$:

$$[\mathbf{U}]\{\mathbf{u}_u\} = \{\mathbf{y}\}$$

Upon solution for $\{\mathbf{u}_u\}$, the reactions that result from deformation of the structure can be found from Eq. (4.44*a*). The total reactions are obtained by subtracting any nodal loads that are applied to the constrained DOF. Such loads frequently exist when element loads, in the form of body forces or surface tractions, are resolved into equivalent nodal loads.

Details of the equation-solving methods and discussion of their advantages and disadvantages can be found in books that specialize in the finite element method (see the Reference section at the end of this chapter).

4.3 BILINEAR RECTANGLE

The constant strain triangle is the simplest element that can be used for plane elasticity problems. As such, it is an attractive choice for demonstration of the basic formulation of the finite element method. However, because of its simplicity, the CST element exhibits relatively poor performance in a coarse mesh (a few large elements). To obtain satisfactory results with the CST element, a very highly refined mesh (many small elements) is generally needed for all but the most trivial problems. Alternatively, one may use a different element that is based on different displacement interpolation functions and that yields better results. The number of alternatives to the CST element is quite large, and no attempt is made to discuss all of them here. Instead, we examine two alternatives: the bilinear rectangle and the linear isoparametric quadrilateral. The development of bilinear rectangle follows. The linear isoparametric quadrilateral is presented in Section 4.4.

Consider a rectangular element of width $2a$, height $2b$, and with corner nodes numbered in counterclockwise order. The (x, y) coordinate axes for the element are parallel to the 1–2 and 1–4 edges of the element, respectively, and the origin of the coordinate system is at the centroid of the element (see Fig. 4.7). As with the CST element, the displacement components (u, v) at any point P are expressed in terms of the nodal displacements. Since there are four nodes in the element, each with two nodal DOFs, the displacement functions for $u(x, y)$ and $v(x, y)$ each have four coefficients. Hence, we choose the bilinear functions[6]

---

[6]These functions are said to be bilinear functions of (x, y) because the dependency on x and y comes from the product of two linear expressions, one in x and one in y. The corresponding rectangular element is said to be bilinear. With the given functions (u, v), the straight edges of the bilinear rectangle remain straight under deformation (like the CST element). However, the strain components in the bilinear rectangle element are not constant.

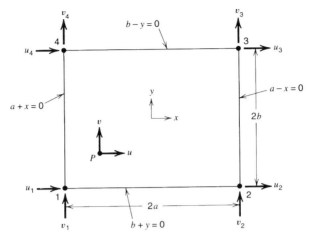

Figure 4.7   Bilinear rectangle element.

$$u(x, y) = a_1 + a_2 x + a_3 y + a_4 xy$$

$$v(x, y) = a_5 + a_6 x + a_7 y + a_8 xy$$

As does the CST element, the *bilinear rectangle* can properly represent rigid-body translation, rigid-body rotation, and constant strain. The bilinear displacement components ($a_4 xy$ and $a_8 xy$) result in strain components such that ϵ_{xx} is linear in y, ϵ_{yy} is linear in x, and γ_{xy} is linear in both x and y. This higher-order response, compared to the CST element, results in more efficient and accurate numerical solutions.

Development of the stiffness matrix and load vector proceeds in a manner similar to that for the CST element. The shape functions are expressed as products of two one-dimensional Lagrange interpolation functions (Kellison, 1975)

$$N_1(x, y) = \frac{(a - x)(b - y)}{4ab}$$

$$N_2(x, y) = \frac{(a + x)(b - y)}{4ab}$$

$$N_3(x, y) = \frac{(a + x)(b + y)}{4ab}$$

$$N_4(x, y) = \frac{(a - x)(b + y)}{4ab}$$

(4.46)

Since the shape function for node i is zero along any element edge that does not include node i, the shape function can be derived directly as the product of the equations of the lines that define these edges (see Fig. 4.7). The shape functions for the bilinear rectangle are illustrated in Fig. 4.8, where they form straight lines along the element edges. However, over the interior of the element, the functions form curved surfaces, with linearly varying slopes in the x and y directions.

The strain–displacement relations are written in the form of Eq. (4.25), with the nodal displacement vector

$$\{u_i\} = [u_1 \quad v_1 \quad u_2 \quad v_2 \quad u_3 \quad v_3 \quad u_4 \quad v_4]^{\mathrm{T}}$$

and $[B]$ matrix

$$[B] = [B_1 \quad B_2 \quad B_3 \quad B_4] \tag{4.47}$$

where for node i

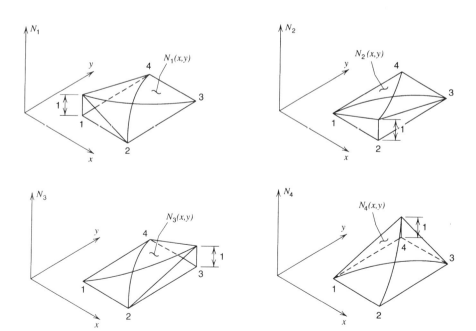

Figure 4.8 Graphical representation of shape functions for the bilinear rectangle element.

$$[B_i] = \begin{bmatrix} \dfrac{\partial N_i}{\partial x} & 0 \\[2mm] 0 & \dfrac{\partial N_i}{\partial y} \\[2mm] \dfrac{\partial N_i}{\partial y} & \dfrac{\partial N_i}{\partial x} \end{bmatrix} \qquad (4.48)$$

The element stiffness matrix is found from Eq. (4.31). That equation is repeated here with the order of each matrix shown as a subscript:

$$[K]_{8 \times 8} = \int_V [B]_{8 \times 3}^{\mathrm{T}} [D]_{3 \times 3} [B]_{3 \times 8} \, dV$$

The stiffness matrix can be written in terms of (2×2) nodal submatrices as

$$[K_{ij}]_{2 \times 2} = \int_V [B_i]_{2 \times 3}^{\mathrm{T}} [D]_{3 \times 3} [B_j]_{3 \times 2} \, dV$$

where i and j are element node numbers. The explicit form of the bilinear rectangle element stiffness matrix for a plane stress condition is given in Table 4.3.

By itself, the bilinear rectangle element is limited to rectangular domains. This is potentially a rather severe restriction. However, non-rectangular domains can be modeled with a combination of bilinear rectangle elements and CST elements. Since both elements represent linear displacement variation along their edges, they are compatible; that is, displacements will be continuous across element boundaries.

Example 4.3: Performance of the Bilinear Rectangle and CST Elements Compare the ability of the bilinear rectangle and the CST elements to model in-plane bending of a thin, square plate.

SOLUTION   A square plate of width a and thickness t is considered. For simplicity, Poisson's ratio is taken as zero, $\nu = 0$. To impose a state of pure bending, displacements $u = \pm \delta$ are imposed on the corners of the plate as shown in Fig. E4.3a. From the theory of elasticity, the displacements are

TABLE 4.3 Bilinear Rectangle Stiffness Matrix, Plane Stress Case (Partitioned into 2 × 2 Nodal Submatrices)

column index $j \rightarrow$

row index $i \downarrow$

	1	2	3	4	5	6	7	8	i
	$4\beta + \dfrac{2(1-\nu)}{\beta}$	$\dfrac{3}{2}(1+\nu)$	$-4\beta + \dfrac{1-\nu}{\beta}$	$-\dfrac{3}{2}(1-3\nu)$	$-2\beta - \dfrac{1-\nu}{\beta}$	$-\dfrac{3}{2}(1+\nu)$	$2\beta - \dfrac{2(1-\nu)}{\beta}$	$\dfrac{3}{2}(1-3\nu)$	1
	$\dfrac{3}{2}(1+\nu)$	$\dfrac{4}{\beta} + 2(1-\nu)\beta$	$\dfrac{3}{2}(1-3\nu)$	$\dfrac{2}{\beta} - 2(1-\nu)\beta$	$-\dfrac{3}{2}(1+\nu)$	$-\dfrac{2}{\beta} - (1-\nu)\beta$	$-\dfrac{3}{2}(1-3\nu)$	$-\dfrac{4}{\beta} + (1-\nu)\beta$	2
	$-4\beta + \dfrac{1-\nu}{\beta}$	$\dfrac{3}{2}(1-3\nu)$	$4\beta + \dfrac{2(1-\nu)}{\beta}$	$-\dfrac{3}{2}(1+\nu)$	$2\beta - \dfrac{2(1-\nu)}{\beta}$	$-\dfrac{3}{2}(1-3\nu)$	$-2\beta - \dfrac{1-\nu}{\beta}$	$\dfrac{3}{2}(1+\nu)$	3
	$-\dfrac{3}{2}(1-3\nu)$	$\dfrac{2}{\beta} - 2(1-\nu)\beta$	$-\dfrac{3}{2}(1+\nu)$	$\dfrac{4}{\beta} + 2(1-\nu)\beta$	$\dfrac{3}{2}(1-3\nu)$	$-\dfrac{4}{\beta} + (1-\nu)\beta$	$\dfrac{3}{2}(1+\nu)$	$-\dfrac{2}{\beta} - (1-\nu)\beta$	4
C	$-2\beta - \dfrac{1-\nu}{\beta}$	$-\dfrac{3}{2}(1+\nu)$	$2\beta - \dfrac{2(1-\nu)}{\beta}$	$\dfrac{3}{2}(1-3\nu)$	$4\beta + \dfrac{2(1-\nu)}{\beta}$	$\dfrac{3}{2}(1+\nu)$	$-4\beta + \dfrac{1-\nu}{\beta}$	$-\dfrac{3}{2}(1-3\nu)$	5
	$-\dfrac{3}{2}(1+\nu)$	$-\dfrac{2}{\beta} - (1-\nu)\beta$	$-\dfrac{3}{2}(1-3\nu)$	$-\dfrac{4}{\beta} + (1-\nu)\beta$	$\dfrac{3}{2}(1+\nu)$	$\dfrac{4}{\beta} + 2(1-\nu)\beta$	$\dfrac{3}{2}(1-3\nu)$	$\dfrac{2}{\beta} - 2(1-\nu)\beta$	6
	$2\beta - \dfrac{2(1-\nu)}{\beta}$	$-\dfrac{3}{2}(1-3\nu)$	$-2\beta - \dfrac{1-\nu}{\beta}$	$\dfrac{3}{2}(1+\nu)$	$-4\beta + \dfrac{1-\nu}{\beta}$	$\dfrac{3}{2}(1-3\nu)$	$4\beta + \dfrac{2(1-\nu)}{\beta}$	$-\dfrac{3}{2}(1+\nu)$	7
	$\dfrac{3}{2}(1-3\nu)$	$-\dfrac{4}{\beta} + (1-\nu)\beta$	$\dfrac{3}{2}(1+\nu)$	$-\dfrac{2}{\beta} - (1-\nu)\beta$	$-\dfrac{3}{2}(1-3\nu)$	$\dfrac{2}{\beta} - 2(1-\nu)\beta$	$-\dfrac{3}{2}(1+\nu)$	$\dfrac{4}{\beta} + 2(1-\nu)\beta$	8

$$C = \frac{Et}{12(1-\nu^2)} \qquad \beta = \frac{b}{a}$$

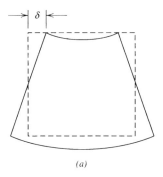

(a)

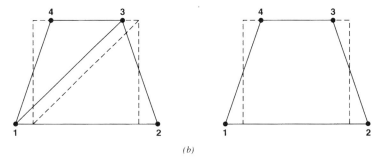

(b)

Figure E4.3   (a) Deformed shape, elasticity solution. (b) Deformed shape, finite element models.

$$u = -\frac{4xy\delta}{a^2} \tag{a}$$

$$v = \left(\frac{4x^2}{a} - 1\right)\frac{\delta}{2} \tag{b}$$

Differentiation of Eqs. (a) and (b) gives the strains

$$\epsilon_{xx} = -\frac{4y\delta}{a^2} \qquad \epsilon_{yy} = 0 \qquad \gamma_{xy} = 0 \tag{c}$$

The strain energy in the plate is

$$U = \int_V U_0 \, dV = \int_V \frac{E\epsilon_{xx}^2}{2} \, dV = \frac{2}{3} Et\delta^2 \tag{d}$$

Two finite element models of the square plate are considered. The first uses two CST elements and the second uses a single bilinear rectangle.

As for the elasticity solution, nodal displacements of $u_i = \pm\delta$ are imposed. The models and their deformed shapes are shown in Fig. E4.3b. The model of two CST elements is considered first. Strains in the CST elements can be determined first, since the displacement vector is known.

$$\{u_i\} = [-\delta \quad 0 \mid \delta \quad 0 \mid -\delta \quad 0 \mid \delta \quad 0]^\mathrm{T} \tag{e}$$

The $[B]$ matrices for the two CST elements are defined by Eq. (4.26). For the element geometries in Fig. E4.3b, these matrices are

$$[B]_{\mathrm{CST}-1} = \frac{1}{a}\begin{bmatrix} 0 & 0 & 1 & 0 & -1 & 0 \\ 0 & -1 & 0 & 0 & 0 & 1 \\ -1 & 0 & 0 & 1 & 1 & -1 \end{bmatrix} \tag{f}$$

$$[B]_{\mathrm{CST}-2} = \frac{1}{a}\begin{bmatrix} -1 & 0 & 1 & 0 & 0 & 0 \\ 0 & 0 & 0 & -1 & 0 & 1 \\ 0 & -1 & -1 & 1 & 1 & 0 \end{bmatrix} \tag{g}$$

Thus, the strains in the two elements are obtained by Eq. (4.25) as

$$\{\epsilon\}_{\mathrm{CST}-1} = \begin{bmatrix} -\dfrac{2\delta}{a} & 0 & \dfrac{2\delta}{a} \end{bmatrix}^\mathrm{T} \tag{h}$$

$$\{\epsilon\}_{\mathrm{CST}-2} = \begin{bmatrix} \dfrac{2\delta}{a} & 0 & -\dfrac{2\delta}{a} \end{bmatrix}^\mathrm{T} \tag{i}$$

The structure stiffness for an assembly of two CST elements is

$$[K]_{\mathrm{CST}} =$$

$$Et\begin{bmatrix} 0.75 & 0.0 & -0.5 & 0.0 & 0.0 & -0.25 & -0.25 & 0.25 \\ 0.0 & 0.75 & 0.25 & -0.25 & -0.25 & 0.0 & 0.0 & -0.5 \\ -0.5 & 0.25 & 0.75 & -0.25 & -0.25 & 0.0 & 0.0 & 0.0 \\ 0.0 & -0.25 & -0.25 & 0.75 & 0.25 & -0.5 & 0.0 & 0.0 \\ 0.0 & -0.25 & -0.25 & 0.25 & 0.75 & 0.0 & -0.5 & 0.0 \\ -0.25 & 0.0 & 0.0 & -0.5 & 0.0 & 0.75 & 0.25 & -0.25 \\ -0.25 & 0.0 & 0.0 & 0.0 & -0.5 & 0.25 & 0.75 & -0.25 \\ 0.25 & -0.5 & 0.0 & 0.0 & 0.0 & -0.25 & -0.25 & 0.75 \end{bmatrix} \tag{j}$$

from which the product $[K]_{\mathrm{CST}}\{u_i\}$ gives the nodal forces $\{P_i\}_{\mathrm{CST}}$ as

$$\{P_i\}_{CST} = Et\delta[-1.5 \quad 0.5 \mid 1.5 - 0.5 \mid -1.5 \quad 0.5 \mid 1.5 - 0.5]^T \quad (k)$$

The strain energy in the structure is

$$U_{CST} = \tfrac{1}{2}\{u_i\}^T\{P_i\}_{CST} = 3Et\delta^2 \tag{l}$$

Next, consider the model with only a single bilinear rectangle shown in Fig. E4.3b. The $[B]$ matrix for the element is given by Eqs. (4.47) and (4.48) as

$$[B]_{BR} =$$

$$\frac{1}{a^2}\begin{bmatrix} -\dfrac{a}{2}+y & 0 & \dfrac{a}{2}-y & 0 & \dfrac{a}{2}+y & 0 & -\dfrac{a}{2}-y & 0 \\[2mm] 0 & -\dfrac{a}{2}+x & 0 & -\dfrac{a}{2}-x & 0 & \dfrac{a}{2}+x & 0 & \dfrac{a}{2}-x \\[2mm] -\dfrac{a}{2}+x & -\dfrac{a}{2}+y & -\dfrac{a}{2}-x & \dfrac{a}{2}-y & \dfrac{a}{2}+x & \dfrac{a}{2}+y & \dfrac{a}{2}-x & -\dfrac{a}{2}-y \end{bmatrix} \quad (m)$$

The strains are obtained by Eq. (4.25) as

$$\{\epsilon\}_{BR} = \left[-\frac{4y\delta}{a^2} \quad 0 \quad -\frac{4x\delta}{a^2}\right]^T \tag{n}$$

The stiffness matrix for the bilinear rectangle is obtained from Table 4.3, which, for this problem, becomes

$$[K]_{BR} = Et$$

$$\begin{bmatrix}
0.5 & 0.125 & -0.25 & -0.125 & -0.25 & -0.125 & 0.0 & 0.125 \\
0.125 & 0.5 & 0.125 & 0.0 & -0.125 & -0.25 & -0.125 & -0.25 \\
-0.25 & 0.125 & 0.5 & -0.125 & 0.0 & -0.125 & -0.25 & 0.125 \\
-0.125 & 0.0 & -0.125 & 0.5 & 0.125 & -0.25 & 0.125 & -0.25 \\
-0.25 & -0.125 & 0.0 & 0.125 & 0.5 & 0.125 & -0.25 & -0.125 \\
-0.125 & -0.25 & -0.125 & -0.25 & 0.125 & 0.5 & 0.125 & 0.0 \\
0.0 & -0.125 & -0.25 & 0.125 & -0.25 & 0.125 & 0.5 & -0.125 \\
0.125 & -0.25 & 0.125 & -0.25 & -0.125 & 0.0 & -0.125 & 0.5
\end{bmatrix}$$

$$(o)$$

from which the product $[K]_{BR}\{u_i\}$ gives the nodal forces $\{P_i\}_{BR}$ as

$$\{P_i\}_{\text{BR}} = Et\delta[-0.5 \quad 0 \mid 0.5 \quad 0 \mid -0.5 \quad 0 \mid 0.5 \quad 0]^{\text{T}} \tag{p}$$

The strain energy in the element is

$$U_{\text{BR}} = \tfrac{1}{2}\{u_i\}^{\text{T}}\{P_i\}_{\text{BR}} = Et\delta^2 \tag{q}$$

This example clearly demonstrates that the bilinear rectangle is superior to the CST element. The bilinear rectangle correctly predicts the normal strains ϵ_{xx} and ϵ_{yy}. In addition, the bilinear rectangle model stores less strain energy than the CST model. Using the elasticity solution as the *exact* solution, $U_{\text{BR}} = 1.5U_{\text{exact}}$, while $U_{\text{CST}} = 4.5U_{\text{exact}}$. Notice, though, that both the CST and bilinear rectangle possess nonzero shear stress where none should exist. This defect, known as *parasitic shear,* contributes to excess strain energy in the elements. Although little can be done to improve the performance of the CST element, a more general formulation of the bilinear rectangle, known as the *linear isoparametric quadrilateral* (Section 4.4), can be used to control parasitic shear.

4.4 LINEAR ISOPARAMETRIC QUADRILATERAL

Suppose that an analyst wishes to model an irregular domain but wants to avoid the use of CST elements because of their relatively poor performance. Since the domain is irregular, the bilinear rectangle element would be inappropriate. Instead, arbitrarily shaped quadrilateral (four-sided) elements are selected to better fit boundaries. A quadrilateral element may be formulated directly, as was done above for the CST and bilinear rectangle elements. However, the necessary integrations are quite complex. This is due, in part, to the difficulty in defining the limits of integration. Use of isoparametric elements eliminates this difficulty. Isoparametric elements are formulated in *natural* coordinates as square elements and then are *mapped* to physical coordinates via coordinate interpolation functions, similar to displacement interpolation functions. Depending on the type of isoparametric element that is used, the configuration of the element in physical coordinates can be non-rectangular and can have curved sides. If the shape functions used for coordinate interpolation are identical to those used for displacement interpolation, the element is said to be *isoparametric.* If coordinate interpolation is of higher order than displacement interpolation (i.e., more nodes are used to represent the variation in geometry than to represent the variation in displacements), the element is called *super-parametric.* If coordinate interpolation is of lower order than displacement interpolation (fewer nodes are used to represent the variation in geometry than to represent the variation in displacements), the element is called *subparametric* (Zienkiewicz and Taylor, 1989, p. 160). Because of their versatility and accuracy, isoparametric elements have become the mainstay of modern finite element programs.

Isoparametric Mapping

Consider the mapping of the four-node quadrilateral element from a *natural* (ξ, η) coordinate system (Fig. 4.9a) to a physical (x, y) coordinate system (Fig. 4.9b). In natural coordinates, the element is a 2×2 square and the origin of the coordinate system is at its center. In physical coordinates, the element is distorted from a rectangular shape. With shape functions in terms of the (ξ, η) coordinate system, the coordinates of any point P can be expressed in terms of the (x, y) coordinates of the nodes.

$$x(\xi, \eta) = \sum_{i=1}^{4} N_i(\xi, \eta)x_i \qquad y(\xi, \eta) = \sum_{i=1}^{4} N_i(\xi, \eta)y_i \quad (4.49a)$$

In matrix form, Eq. (4.49a) is

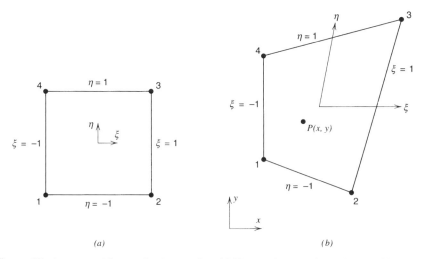

(a)   (b)

Figure 4.9 Isoparametric coordinate mapping. (a) Element in natural coordinates. (b) Element in physical coordinates.

$$\begin{Bmatrix} x(\xi, \eta) \\ y(\xi, \eta) \end{Bmatrix} = [N]\{x_i\} \qquad (4.49b)$$

where $\{x_i\}$ is the vector of nodal coordinates

$$\{x_i\} = [x_1 \quad y_1 \quad x_2 \quad y_2 \quad x_3 \quad y_3 \quad x_4 \quad y_4]^{\mathrm{T}}$$

and $[N]$ is the shape function matrix

$$[N] = \begin{bmatrix} N_1 & 0 & N_2 & 0 & N_3 & 0 & N_4 & 0 \\ 0 & N_1 & 0 & N_2 & 0 & N_3 & 0 & N_4 \end{bmatrix} \qquad (4.50)$$

The shape functions are the Lagrange interpolation functions [refer to Eq. (4.46)] in dimensionless (ξ, η) coordinates

$$N_1(\xi, \eta) = \frac{(1 - \xi)(1 - \eta)}{4}$$

$$N_2(\xi, \eta) = \frac{(1 + \xi)(1 - \eta)}{4}$$

$$N_3(\xi, \eta) = \frac{(1 + \xi)(1 + \eta)}{4} \qquad (4.51)$$

$$N_4(\xi, \eta) = \frac{(1 - \xi)(1 + \eta)}{4}$$

After the element is mapped from natural to physical coordinates, the ξ- and η-axes need not remain orthogonal.

The principal reason for using isoparametric elements is to avoid integrating in physical coordinates. However, the general expression for the stiffness matrix, Eq. (4.31), is expressed in terms of physical coordinates. Therefore, the differential lengths dx and dy must be expressed in terms of the natural coordinate differentials $d\xi$ and $d\eta$. In addition, strain is defined in terms of the derivatives of the shape functions with respect to physical coordinates. These derivatives are the elements in the $[B]$ matrix, and they must be converted to derivatives with respect to natural coordinates.

The differentials (dx, dy) are related to the differentials ($d\xi, d\eta$) by means of Eq. (4.49a). Thus

$$dx = \frac{\partial x}{\partial \xi} d\xi + \frac{\partial x}{\partial \eta} d\eta$$

$$dy = \frac{\partial y}{\partial \xi} d\xi + \frac{\partial y}{\partial \eta} d\eta$$

(4.52)

where

$$\frac{\partial x}{\partial \xi} = \sum \frac{\partial N_i}{\partial \xi} x_i \qquad \frac{\partial y}{\partial \xi} = \sum \frac{\partial N_i}{\partial \xi} y_i$$

$$\frac{\partial x}{\partial \eta} = \sum \frac{\partial N_i}{\partial \eta} x_i \qquad \frac{\partial y}{\partial \eta} = \sum \frac{\partial N_i}{\partial \eta} y_i$$

The coordinate derivatives are combined in matrix form as

$$[J] = \begin{bmatrix} \dfrac{\partial x}{\partial \xi} & \dfrac{\partial y}{\partial \xi} \\ \dfrac{\partial x}{\partial \eta} & \dfrac{\partial y}{\partial \eta} \end{bmatrix}$$

(4.53)

where $[J]$ is the *Jacobian* of the transformation (Courant, 1950).

Equations (4.52) and (4.53) relate the differentials of the two coordinate systems as

$$\begin{Bmatrix} dx \\ dy \end{Bmatrix} = [J]^{\mathrm{T}} \begin{Bmatrix} d\xi \\ d\eta \end{Bmatrix}$$

(4.54)

In a similar manner, derivatives of the shape function for node i are related by

$$\begin{Bmatrix} \dfrac{\partial N_i}{\partial x} \\ \dfrac{\partial N_i}{\partial y} \end{Bmatrix} = [J]^{-1} \begin{Bmatrix} \dfrac{\partial N_i}{\partial \xi} \\ \dfrac{\partial N_i}{\partial \eta} \end{Bmatrix}$$

(4.55)

If $[J]^{-1}$ exists, the area mapping from (ξ, η) coordinates to (x, y) coordinates is unique and reversible. A physical interpretation of $[J]$ can be obtained by comparing the area of the element in (x, y) coordinates to that in (ξ, η) coordinates. If $|J| > 0$, the area of the element is preserved and the mapping is physically meaningful. In precise terms, $|J|$ is the instantaneous area ratio $A_{xy}/A_{\xi\eta}$ at any point in the element.

This physical interpretation of $[J]$ leads to a change in the differential volume for a constant-thickness plane elasticity element from $t\,dx\,dy$ to $t|J|\,d\xi\,d\eta$. The limits of integration are -1 to 1 in ξ and -1 to 1 in η. So the integral of any function $F(x, y)$ can be transformed to natural coordinates in the manner

$$\int_A F(x, y)\,dx\,dy = \int_{-1}^{1}\int_{-1}^{1} F(x(\xi, \eta), y(\xi, \eta))|J|\,d\xi\,d\eta$$

Element Stiffness Matrix

Equation (4.31) defines the element stiffness matrix for any elasticity element (using displacement DOFs), including the isoparametric linear quadrilateral. A change in coordinate system from (x, y) to (ξ, η), with the modified limits of integration, leads to the stiffness matrix

$$[K] = t \int_{-1}^{1}\int_{-1}^{1} [B]^{\mathrm{T}}[D][B]|J|\,d\xi\,d\eta \qquad (4.56)$$

where $[B]$ is given by Eq. (4.47) and $[B_i]$ by Eq. (4.48). From Eqs. (4.48) and (4.55), the individual terms in $[B_i]$, in terms of (ξ, η), are

$$[B_i(\xi, \eta)] = \begin{bmatrix} J_{11}^{*}\dfrac{\partial N_i}{\partial \xi} + J_{12}^{*}\dfrac{\partial N_i}{\partial \eta} & 0 \\[2ex] 0 & J_{21}^{*}\dfrac{\partial N_i}{\partial \xi} + J_{22}^{*}\dfrac{\partial N_i}{\partial \eta} \\[2ex] J_{21}^{*}\dfrac{\partial N_i}{\partial \xi} + J_{22}^{*}\dfrac{\partial N_i}{\partial \eta} & J_{11}^{*}\dfrac{\partial N_i}{\partial \xi} + J_{12}^{*}\dfrac{\partial N_i}{\partial \eta} \end{bmatrix} \qquad (4.57)$$

where J_{ij}^{*} is the i, j term from $[J]^{-1}$.

It is usually more convenient to work with just a single (2×2) nodal submatrix of $[K]$ at one time. So we write

$$[K_{ij}] = t \int_{-1}^{1}\int_{-1}^{1} [B_i]^{\mathrm{T}}[D][B_j]|J|\,d\xi\,d\eta \qquad (4.58)$$

where i and j are node numbers for the element.

Numerical Integration

While analytical expressions for the individual terms in Eq. (4.58) can be developed, they are quite complex and thus prone to errors in algebra or in computer programming. As an alternative to direct integration, the integrals required are usually evaluated numerically within the finite element program. The most commonly used numerical integration method is *Gauss quadrature*. The Gauss quadrature method is more efficient than many other methods, such as the Newton–Cotes methods, since fewer sampling points are required to obtain a given accuracy. In fact, in one dimension, the use of n sampling points in Gauss quadrature results in exact integration of a polynomial of order $(2n - 1)$. However, the integration of functions that are not polynomials is approximate.

Consider a function $F(\xi, \eta)$ that is to be integrated over the limits of -1 to 1 in ξ and -1 to 1 in η. The integral is evaluated numerically by the form

$$I = \int_{-1}^{1} \int_{-1}^{1} F(\xi, \eta) \, d\xi \, d\eta = \sum_{k=1}^{m} \sum_{l=1}^{n} w_k w_l F(\xi_k, \eta_l)$$

where m and n are the number of sampling points in the ξ and η directions, respectively. Also, ξ_k and η_l are the locations of the kth and lth sampling points, and w_k and w_l are weights applied to $F(\xi, \eta)$ when it is evaluated at the sampling points. Usually, m and n are taken equal, in which case the numerical scheme is symmetric.

If Gauss quadrature is used to evaluate the nodal submatrix $[K_{ij}]$ in Eq. (4.58), the integral becomes

$$[K_{ij}] = t \sum_{k=1}^{m} \sum_{l=1}^{n} w_k w_l [B_i(\xi_k, \eta_l)]^{\mathrm{T}} [D][B_j(\xi_k, \eta_l)] |J(\xi_k, \eta_l)| \quad (4.59)$$

The accuracy achieved with Gauss quadrature is dependent on proper selection of sampling point locations and weights. For elements in natural coordinates, the optimal sampling point locations and weights are given in Fig. 4.10. Only symmetric integration and the one-, two-, and three-point rules are considered. Nonsymmetric integration and higher-order integration rules are discussed elsewhere.

The number of integration points that are used to evaluate Eq. (4.59) influences the ultimate performance of the element. *Full integration* is the integration order needed to integrate the stiffness exactly for an

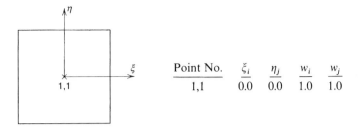

Point No.	ξ_i	η_j	w_i	w_j
1,1	0.0	0.0	1.0	1.0

(a)

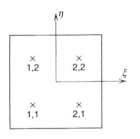

Point No.	ξ_i	η_j	w_i	w_j
1,1	$-1/\sqrt{3}$	$-1/\sqrt{3}$	1.0	1.0
2,1	$1/\sqrt{3}$	$-1/\sqrt{3}$	1.0	1.0
1,2	$-1/\sqrt{3}$	$1/\sqrt{3}$	1.0	1.0
2,2	$1/\sqrt{3}$	$1/\sqrt{3}$	1.0	1.0

(b)

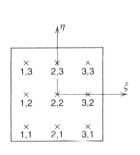

Point No.	ξ_i	η_j	w_i	w_j
1,1	$-\sqrt{0.6}$	$-\sqrt{0.6}$	5/9	5/9
2,1	0	$-\sqrt{0.6}$	8/9	5/9
3,1	$\sqrt{0.6}$	$-\sqrt{0.6}$	5/9	5/9
1,2	$-\sqrt{0.6}$	0	5/9	8/9
2,2	0	0	8/9	8/9
3,2	$\sqrt{0.6}$	0	5/9	8/9
1,3	$-\sqrt{0.6}$	$\sqrt{0.6}$	5/9	5/9
2,3	0	$\sqrt{0.6}$	8/9	5/9
3,3	$\sqrt{0.6}$	$\sqrt{0.6}$	5/9	5/9

(c)

Figure 4.10 Optimal sampling point locations and weights for Gauss quadrature. (a) One-point rule. (b) Two-point rule. (c) Three-point rule.

undistorted element. For the linear quadrilateral, a two-point rule provides full integration. An integration rule below that required for full integration is termed *reduced integration*. Reduced integration, although not evaluating Eq. (4.59) exactly, can often lead to improved performance of an element, relative to full integration. For instance, reduced integration of the linear quadrilateral can eliminate the parasitic shear that is a common defect in the element (see Example 4.3). A more complete discussion of reduced integration, including justification for its use, can be found in most finite element textbooks.

High-order Isoparametric Elements

The concept of isoparametric mapping has been applied to a broad list of element geometries. Within the scope of plane elasticity problems, elements with more than four nodes permit greater flexibility in element shape (including curved edges) and are capable of representing greater variation in displacements. Perhaps the most popular of all isoparametric elements is the eight-node quadrilateral. This element has four corner nodes, like the linear quadrilateral, but it also has four midside nodes, one midway along the length of each edge (Fig. 4.11*a*). With three nodes along each edge, the element can have curved (parabolic) sides. Another popular high-order isoparametric element is the nine-

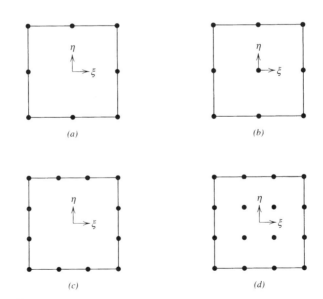

(a) *(b)*

(c) *(d)*

Figure 4.11 Higher-order isoparametric elements. (*a*) Quadratic serendipity element. (*b*) Quadratic Lagrange element. (*c*) Cubic serendipity element. (*d*) Cubic Lagrange element.

node quadrilateral (Fig. 4.11*b*). This element has four corner nodes, four mid-side nodes, and one interior node. Both the eight- and nine-node elements represent complete quadratic displacement fields. The generalization of these elements to cubic interpolation is straightforward (see Fig. 4.11*c* and *d*).

The eight-node quadrilateral and other high-order elements that contain only boundary nodes are known as *serendipity elements*. The term *serendipity* is used because shape functions for this family of elements were initially developed by inspection. The nine-node quadrilateral and other high-order elements that contain a regular pattern of nodes are known as *Lagrangian elements* since their shape functions are based on Lagrange interpolation functions.

4.5   PLANE FRAME ELEMENT

Analysis of framed structures by the *stiffness method* (also known as *matrix analysis*) was fairly well established at the time of development of the finite element method. The stiffness method for frame analysis can be developed entirely from basic mechanics of materials principles, without the need to consider virtual work formulations and interpolation polynomials. As a result, many engineers view the two methods as distinct. However, it is clear that the stiffness method for frames is simply a special case of the finite element method. Hence, in this section we develop a finite element that represents a plane frame member, using the same approach that was used for plane elasticity problems.

Element Stiffness Matrix

The classical plane frame element has two nodes, it is straight and prismatic, and it has three DOFs and three corresponding end actions at each node (see Fig. 4.12*a*). The element has cross-sectional area A, moment of inertia I, and modulus of elasticity E. We assume that the axial response of the member is independent of the bending response. Consequently, the frame element stiffness is formulated as a superposition of the stiffness for an axial rod and that for a beam (Fig. 4.12*b*). In the following, a local $(\bar{x}, \bar{y})$ coordinate system is established for the element. The local $\bar{x}$-axis is aligned with the longitudinal axis of the member and the $\bar{y}$-axis lies in the plane of the element cross section. The stiffness matrix for the frame element is derived in terms of this local coordinate system. When the element is oriented at some angle

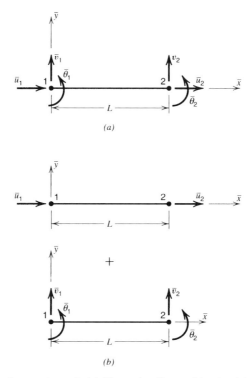

Figure 4.12 Plane frame element. (a) Element with combined axial and bending DOFs. (b) Axial and bending DOFs treated separately.

ϕ with respect to the *global* (x, y) coordinates for the structure, the nodal DOFs of the element must be related to the global coordinate system. Thus, a coordinate rotation from local to global coordinates is required for the displacements, loads, and stiffness. This rotation is discussed following Eq. (4.75).

Consider first the case of the axial rod. There are two nodal DOFs associated with axial response, so the displacement variation is taken as a linear function:

$$\bar{u}(x) = a_0 + a_1 \bar{x}$$

The coefficients a_0 and a_1 are evaluated based on the boundary conditions $\bar{u}(0) = \bar{u}_1$ and $\bar{u}(L) = \bar{u}_2$, where L is the element length. The displacement function, in terms of the nodal displacements, becomes

$$\bar{u}(x) = [N]\{\bar{u}_i\}$$

where $[N] = [1 - \bar{x}L \quad \bar{x}/L]$ and $\{\bar{u}_i\} = [\bar{u}_1 \quad \bar{u}_2]^{\mathrm{T}}$.

The only nonzero strain component is ϵ_{xx}, which is written in terms of the nodal displacements as

$$\epsilon_{xx} = [B_A]\{\bar{u}_i\}$$

in which the subscript A indicates *axial* response and

$$[B_A] = \begin{bmatrix} \dfrac{\partial N_1}{\partial \bar{x}} & \dfrac{\partial N_2}{\partial \bar{x}} \end{bmatrix} = \begin{bmatrix} -\dfrac{1}{L} & \dfrac{1}{L} \end{bmatrix}$$

The axial stress is written as $\sigma_{xx} = E\epsilon_{xx}$ and the variation of internal energy is

$$\delta U = \int_V \delta\epsilon_{xx}\sigma_{xx}\, dV \tag{4.60}$$

Assume that only concentrated nodal loads are applied. Then, substitution for σ_{xx} and $\delta\epsilon_{xx}$ in Eq. (4.60), and then substitution of Eq. (4.60) into (4.1) yields

$$\{\delta\bar{u}_i\}^{\mathrm{T}}\{\bar{F}_i\} - \{\delta\bar{u}_i\}^{\mathrm{T}}\left[A \int_0^L [B_A]^{\mathrm{T}}E[B_A]\, d\bar{x} \right]\{\bar{u}_i\} = 0$$

Since $\{\delta\bar{u}_i\}$ is arbitrary,

$$\{\bar{F}_i\} = \left[A \int_0^L [B_A]^{\mathrm{T}}E[B_A]\, d\bar{x} \right]\{\bar{u}_i\}$$

which leads to the stiffness matrix for the axial rod:

$$[\bar{K}_A] = A \int_0^L [B_A]^{\mathrm{T}}E[B_A]\, d\bar{x}$$

For constant E, the integrals are easily evaluated to obtain $[\bar{K}_A]$ in terms of A, E, and L:

$$[\bar{K}_A] = \begin{bmatrix} \dfrac{AE}{L} & -\dfrac{AE}{L} \\ -\dfrac{AE}{L} & \dfrac{AE}{L} \end{bmatrix} \tag{4.61}$$

Next consider the *bending* effect of the frame element. There are four nodal DOFs associated with bending (a lateral translation and a rotation at each node), so the displacement variation is written as a cubic polynomial with four coefficients:

$$\bar{u}(x) = a_0 + a_1\bar{x} + a_2\bar{x}^2 + a_3\bar{x}^3$$

Coefficients a_0 through a_3 are evaluated based on the boundary conditions $\bar{v}(0) = \bar{v}_1$, $\bar{\theta}(0) = \bar{\theta}_1$, $\bar{v}(L) = \bar{v}_2$, and $\bar{\theta}(L) = \bar{\theta}_2$ in which $\bar{\theta} = d\bar{v}/d\bar{x}$. In terms of the nodal displacements, the displacement function is

$$\bar{v}(x) = [N]\{\bar{v}_i\} \tag{4.62}$$

where $\{\bar{v}_i\} = [\bar{v}_1 \quad \bar{\theta}_1 \quad \bar{v}_2 \quad \bar{\theta}_2]^{\mathrm{T}}$, and the shape function matrix $[N]$ is

$$[N] = [N_1 \quad N_2 \quad N_3 \quad N_4] \tag{4.63a}$$

for which the individual shape functions are

$$N_1 = 1 - 3\frac{\bar{x}^2}{L^2} + 2\frac{\bar{x}^3}{L^3}$$

$$N_2 = \bar{x} - 2\frac{\bar{x}^2}{L} + \frac{\bar{x}^3}{L^2}$$

$$N_3 = 3\frac{\bar{x}^2}{L^2} - 2\frac{\bar{x}^3}{L^3} \tag{4.63b}$$

$$N_4 = -\frac{\bar{x}^2}{L} + \frac{\bar{x}^3}{L^2}$$

These shape functions are illustrated in Fig. 4.13.

The strain energy in a beam subjected to bending is given by

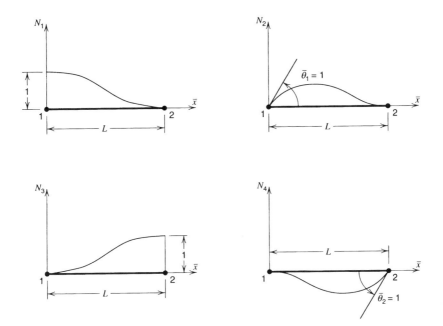

Figure 4.13   Beam element shape functions.

$$U = \int_0^L \frac{M^2}{2EI} \, d\bar{x} \tag{4.63c}$$

If the curvature $\bar{v}''$ is taken as a *generalized strain* quantity, the strain-nodal displacement relation is

$$\bar{v}''(\bar{x}) = [B_B]\{\bar{v}_i\} \tag{4.64a}$$

where the subscript B represents *bending* response and where

$$[B_B] = \left[\frac{d^2N_1}{d\bar{x}^2} \quad \frac{d^2N_2}{d\bar{x}^2} \quad \frac{d^2N_3}{d\bar{x}^2} \quad \frac{d^2N_4}{d\bar{x}^2} \right] \tag{4.64b}$$

Substitution of $M = EI\bar{v}''$ into Eq. (4.63c) gives

$$U = \int_0^L \frac{EI(\bar{v}'')^2}{2} \, d\bar{x} \tag{4.65}$$

from which the first variation of the strain energy is

$$\delta U = \int_0^L (\delta\bar{v})'' EI\bar{v}'' \, d\bar{x}$$

In terms of nodal DOFs, from Eq. (4.64a), δU is

$$\delta U = \int_0^L \{\delta\bar{v}_i\}^{\mathrm{T}} [B_B]^{\mathrm{T}} EI[B_B]\{\bar{v}_i\} \, d\bar{x} \tag{4.66}$$

In the manner followed with other elements, only nodal loads are assumed, Eq. (4.66) is substituted into Eq. (4.1), $\{\delta v_i\}$ is eliminated, and the bending stiffness matrix is found to be

$$[\bar{K}_B] = \int_0^L [B_B]^{\mathrm{T}} EI[B_B] \, d\bar{x}$$

Since EI is constant, integration yields the bending stiffness matrix in terms of E, I, and L as

$$[\bar{K}_B] = \begin{bmatrix} \dfrac{12EI}{L^3} & \dfrac{6EI}{L^2} & -\dfrac{12EI}{L^3} & \dfrac{6EI}{L^2} \\[2mm] \dfrac{6EI}{L^2} & \dfrac{4EI}{L} & -\dfrac{6EI}{L^2} & \dfrac{2EI}{L} \\[2mm] -\dfrac{12EI}{L^3} & -\dfrac{6EI}{L^2} & \dfrac{12EI}{L^3} & -\dfrac{6EI}{L^2} \\[2mm] \dfrac{6EI}{L^2} & \dfrac{2EI}{L} & -\dfrac{6EI}{L^2} & \dfrac{4EI}{L} \end{bmatrix} \tag{4.67}$$

The stiffness matrix for the plane frame element [see Eq. (4.68)] is a combination of the axial stiffness matrix, Eq. (4.61), and the bending stiffness matrix, Eq. (4.67). Note that the ordering of the DOFs in the element first lists all three DOFs at node 1 and then the three DOFs at node 2.

$$[\bar{K}] = \begin{bmatrix} \dfrac{AE}{L} & 0 & 0 & \dfrac{-AE}{L} & 0 & 0 \\[2mm] 0 & \dfrac{12EI}{L^3} & \dfrac{6EI}{L^2} & 0 & \dfrac{-12EI}{L^3} & \dfrac{6EI}{L^2} \\[2mm] 0 & \dfrac{6EI}{L^2} & \dfrac{4EI}{L} & 0 & \dfrac{-6EI}{L^2} & \dfrac{2EI}{L} \\[2mm] \dfrac{-AE}{L} & 0 & 0 & \dfrac{AE}{L} & 0 & 0 \\[2mm] 0 & \dfrac{-12EI}{L^3} & \dfrac{-6EI}{L^2} & 0 & \dfrac{12EI}{L^3} & \dfrac{-6EI}{L^2} \\[2mm] 0 & \dfrac{6EI}{L^2} & \dfrac{2EI}{L} & 0 & \dfrac{-6EI}{L^2} & \dfrac{4EI}{L} \end{bmatrix} \quad (4.68)$$

The displacement vector $\{\bar{u}_i\}$ for the element is

$$\{\bar{u}_i\} = [\bar{u}_1 \quad \bar{v}_1 \quad \bar{\theta}_1 \quad \bar{u}_2 \quad \bar{v}_2 \quad \bar{\theta}_2]^{\mathrm{T}} \quad (4.69)$$

and the element end action (load) vector $\{\bar{P}_i\}$ is

$$\{\bar{P}_i\} = [\bar{P}_{x1} \quad \bar{P}_{y1} \quad \bar{M}_1 \quad \bar{P}_{x2} \quad \bar{P}_{y2} \quad \bar{M}_2]^{\mathrm{T}} \quad (4.70)$$

Finally, the relationship between nodal loads and nodal displacements for an element in local coordinates is given by the familiar form

$$[\bar{K}]\{\bar{u}_i\} = \{\bar{P}_i\} \quad (4.71)$$

Equivalent Nodal Load Vector

As for most other elements, actual loads that are applied over the element must be converted to equivalent nodal loads. We consider only element loads that affect beam behavior. Two cases are considered: a distributed load over a portion of the element and a transverse concentrated force. Equivalent nodal loads for axial behavior are derived in a similar fashion.

For a distributed load along the beam, not necessarily over the full length, the variation of work δW_D of the load is

$$\delta W_D = \int_{L_a}^{L_b} \delta \bar{v} \, \bar{q}(\bar{x}) \, d\bar{x} \tag{4.72}$$

where $\bar{q}(\bar{x})$ is the load function that exists over the domain $L_a < \bar{x} < L_b$ (see Fig. 4.14a) and the subscript D denotes a distributed load. Equation (4.62) is substituted into (4.72) and the equivalent nodal load vector is obtained as

$$\{\bar{P}_{Di}\} = \int_{L_a}^{L_b} [N]^T \bar{q}(\bar{x}) \, d\bar{x} \tag{4.73}$$

For a concentrated load $\bar{P}_C$ located at $\bar{x} = L_c$ along the beam (see Fig. 4.14b) the variation of work δW_C of the load is

$$\delta W_C = \delta \bar{v}|_{\bar{x}=L_c} \bar{P}_C \tag{4.74}$$

variation of displacement $\delta \bar{v}$ at $\bar{x} = L_c$ is written in terms of the variation of nodal displacements by Eq. (4.62) with the shape functions evaluated at $\bar{x} = L_c$. The equivalent nodal load vector is

$$\{\bar{P}_{Ci}\} = [N]|_{\bar{x}=L_c}^T \bar{P}_C \tag{4.75}$$

By Eqs. (4.73) and (4.75), equivalent nodal load vectors for several load patterns on a beam element were determined and are shown in Fig. 4.15.

Coordinate Rotations

Consider an element in a structure oriented at an angle ϕ with respect to the global x-axis (Fig. 4.16). To assemble the stiffness matrix and load vector for this element with those of other elements, all nodal

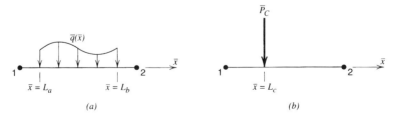

Figure 4.14 Element loads for beam element. (a) Distributed load. (b) Concentrated load.

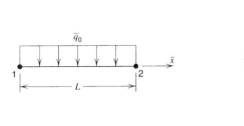

$$\{\bar{P}_{Di}\} = \begin{Bmatrix} -\dfrac{\bar{q}_0 L}{2} \\[2mm] -\dfrac{\bar{q}_0 L^2}{12} \\ \hline -\dfrac{\bar{q}_0 L}{2} \\[2mm] \dfrac{\bar{q}_0 L^2}{12} \end{Bmatrix}$$

(a)

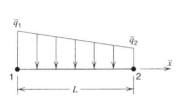

$$\{\bar{P}_{Di}\} = \begin{Bmatrix} -\dfrac{L}{20}(7\bar{q}_1 + 3\bar{q}_2) \\[2mm] -\dfrac{L^2}{60}(3\bar{q}_1 + 2\bar{q}_2) \\ \hline -\dfrac{L}{60}(3\bar{q}_1 + 7\bar{q}_2) \\[2mm] \dfrac{L^2}{60}(2\bar{q}_1 + 3\bar{q}_2) \end{Bmatrix}$$

(b)

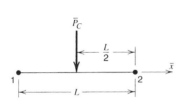

$$\{\bar{P}_{Ci}\} = \begin{Bmatrix} -\dfrac{\bar{P}_C}{2} \\[2mm] -\dfrac{\bar{P}_C L}{8} \\ \hline -\dfrac{\bar{P}_C}{2} \\[2mm] \dfrac{\bar{P}_C L}{8} \end{Bmatrix}$$

(c)

Figure 4.15   Equivalent nodal loads for beam element. (a) Uniformly distributed load. (b) Linearly distributed load. (c) Concentrated load.

DOFs must be defined in terms of the global coordinate system. For node i, the displacements in the two coordinate systems are related by

$$\begin{Bmatrix} \bar{u}_i \\ \bar{v}_i \\ \bar{\theta}_i \end{Bmatrix} = [\lambda] \begin{Bmatrix} u_i \\ v_i \\ \theta_i \end{Bmatrix} \tag{4.76}$$

where

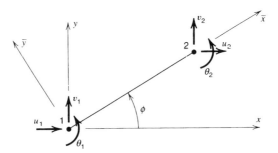

Figure 4.16   Frame element in global coordinates.

$$[\lambda] = \begin{bmatrix} \cos\phi & \sin\phi & 0 \\ -\sin\phi & \cos\phi & 0 \\ 0 & 0 & 1 \end{bmatrix}$$

For a plane frame element, with two nodes, the displacements are related by

$$\{\bar{u}_i\} = [T]\{u_i\} \tag{4.77}$$

where the rotation (transformation) matrix $[T]$ is

$$[T] = \begin{bmatrix} \lambda & 0 \\ 0 & \lambda \end{bmatrix}$$

In a similar manner, element end actions (loads) are rotated by

$$\{\bar{P}_i\} = [T]\{P_i\} \tag{4.78}$$

Substitution of Eqs. (4.77) and (4.78) into (4.71) yields

$$[\bar{K}][T]\{u_i\} = [T]\{P_i\} \tag{4.79}$$

Premultipling both sides of Eq. (4.79) by $[T]^{-1}$ and observing that $[T]^{-1} = [T]^T$, since $[T]$ is an orthogonal matrix, we obtain

$$[T]^T[\bar{K}][T]\{u_i\} = \{P_i\}$$

Thus, since $\{u_i\}$ and $\{P_i\}$ are in global coordinates, the stiffness matrix for the plane frame element, in global coordinates, is

$$[K] = [T]^\mathrm{T}[\overline{K}][T] \tag{4.80}$$

The final form of $[K]$ is given in Table 4.4. The load vector for the element, in global coordinates, is obtained from Eq. (4.78) as

$$\{P_i\} = [T]^\mathrm{T}\{\overline{P}_i\} \tag{4.81}$$

4.6   CLOSING REMARKS

Requirements for Accuracy

The accuracy of a finite element solution depends strongly on two conditions. First, it is important that the equations of equilibrium be satisfied throughout the model. Second, it is important that compatibility (continuity of displacements) be maintained. In certain circumstances, these conditions are violated, as noted below.

Equilibrium at the structure nodes is satisfied since the basic system of equations, (4.45), is fundamentally a system of nodal equilibrium equations. Thus, within the accuracy of the equation-solving process (numerical error), the structure nodes are in equilibrium.

For elements with only displacement DOFs, equilibrium along element edges is generally not satisfied. This is due to the fact that while displacements might be continuous across element boundaries, their derivatives are not, and thus stresses are not continuous. For instance, consider two constant-strain triangle elements, such as those shown in Fig. 4.17. Nodes 1, 2, and 3 are fully constrained while node 4 has an imposed displacement in the x direction. Hence, element 1 is unstressed while element 2 has nonzero σ_{xx}. Due to the stress discontinuity, a differential element located at the boundary between the two elements does not satisfy equilibrium in the x direction.

Equilibrium within an element is not satisfied unless body forces are of relatively low order or are entirely absent. For a constant-strain triangle, the stress state is constant throughout the element. Thus, equilibrium of a differential element is satisfied only when body forces are absent. Similarly, for elements that can represent linear stress variation, body forces must be, at most, constant in magnitude for equilibrium.

Compatibility at the nodes is assured due to the assembly process. That is, the displacements of adjacent elements are the same at their common nodes. However, to assure that compatibility is maintained along the common edge between two adjacent elements, the displacements along that edge, viewed from either element, must be expressed

TABLE 4.4 Element Stiffness Matrix for Plane Frame Element in Global Coordinates

$$[K] = \begin{bmatrix}
c^2\dfrac{AE}{L} + s^2\dfrac{12EI}{L^3} & sc\left(\dfrac{AE}{L} - \dfrac{12EI}{L^3}\right) & -s\left(\dfrac{6EI}{L^2}\right) & -c^2\dfrac{AE}{L} - s^2\dfrac{12EI}{L^3} & -sc\left(\dfrac{AE}{L} - \dfrac{12EI}{L^3}\right) & -s\left(\dfrac{6EI}{L^2}\right) \\[2ex]
sc\left(\dfrac{AE}{L} - \dfrac{12EI}{L^3}\right) & s^2\dfrac{AE}{L} + c^2\dfrac{12EI}{L^3} & c\left(\dfrac{6EI}{L^2}\right) & -sc\left(\dfrac{AE}{L} - \dfrac{12EI}{L^3}\right) & -s^2\dfrac{AE}{L} - c^2\dfrac{12EI}{L^3} & c\left(\dfrac{6EI}{L^2}\right) \\[2ex]
-s\left(\dfrac{6EI}{L^2}\right) & c\left(\dfrac{6EI}{L^2}\right) & \dfrac{4EI}{L} & s\left(\dfrac{6EI}{L^2}\right) & -c\left(\dfrac{6EI}{L^2}\right) & \dfrac{2EI}{L} \\[2ex]
-c^2\dfrac{AE}{L} - s^2\dfrac{12EI}{L^3} & -sc\left(\dfrac{AE}{L} - \dfrac{12EI}{L^3}\right) & s\left(\dfrac{6EI}{L^2}\right) & c^2\dfrac{AE}{L} + s^2\dfrac{12EI}{L^3} & sc\left(\dfrac{AE}{L} - \dfrac{12EI}{L^3}\right) & s\left(\dfrac{6EI}{L^2}\right) \\[2ex]
-sc\left(\dfrac{AE}{L} - \dfrac{12EI}{L^3}\right) & -s^2\dfrac{AE}{L} - c^2\dfrac{12EI}{L^3} & -c\left(\dfrac{6EI}{L^2}\right) & sc\left(\dfrac{AE}{L} - \dfrac{12EI}{L^3}\right) & s^2\dfrac{AE}{L} + c^2\dfrac{12EI}{L^3} & -c\left(\dfrac{6EI}{L^2}\right) \\[2ex]
-s\left(\dfrac{6EI}{L^2}\right) & c\left(\dfrac{6EI}{L^2}\right) & \dfrac{2EI}{L} & s\left(\dfrac{6EI}{L^2}\right) & -c\left(\dfrac{6EI}{L^2}\right) & \dfrac{4EI}{L}
\end{bmatrix}$$

column index $j \rightarrow$ 1, 2, 3, 4, 5, 6

row index $i \downarrow$ 1, 2, 3, 4, 5, 6

$$c = \cos\phi \qquad s = \sin\phi$$

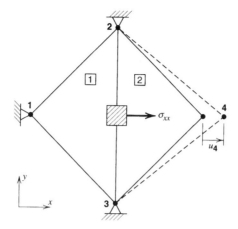

Figure 4.17 Equilibrium along element edges.

entirely in terms of the displacement of nodes on that edge. Elements that maintain compatibility along common edges are known as *conforming* elements. Generally, this condition is satisfied for elements that possess only translational DOFs. However, certain plate-bending and shell elements, for instance, are nonconforming. Compatibility within an element is assured as long as the displacement interpolation polynomials are continuous.

Requirements for Convergence

As discussed at the beginning of this chapter, a major source of error in a finite element solution is due to the use of approximation functions to describe element response (formulation error). To reduce formulation error, we successively refine our finite element models with the expectation that the numerical solution will converge to the *exact* solution. Under certain conditions, convergence can be guaranteed. These conditions are:

1. *The elements must be complete.* That is, the shape function must be a complete polynomial. For instance, a complete quadratic contains all possible quadratic terms and omits no linear or constant terms. Inclusion of a few cubic terms, such as for the quadratic serendipity and Lagrange elements, does not destroy completeness of the quadratic polynomial.
2. *The elements must be compatible.* Hence, continuity of displacements must be assured throughout the entire structural model.

3. *The elements must be capable of representing rigid-body motion and constant strain.* For two- and three-dimensional elasticity problems, these are assured if the displacement field contains at least a complete linear polynomial. For shell elements, constant strain implies constant curvature w_{xx} and w_{yy} and constant twist w_{xy}. Some shell elements have difficulty representing rigid-body motion.

Generally, a finite element model is too stiff. That is, displacements converge from below. A qualitative explanation is as follows. The elements are *constrained,* by the shape functions, to deform in a specific (unnatural) manner. This constraint adds stiffness, relative to the physical system, that results in smaller displacements when the external influences on the system are loads. If all external loads are zero and the only external influences on the system are imposed (nonzero) displacements, additional energy is required to force the model into the imposed deformed shape.

In the case of isoparametric elements, reduced integration can be used effectively to *soften* the element such that its response improves relative to full integration. Problem 4.1 demonstrates how the use of approximation functions to represent displacements results in a model that is stiff relative to the actual system.

Modeling Recommendations

As an aid to the application of the finite element method to analysis of practical problems in elasticity, the following recommendations are offered. The list is not exhaustive and the recommendations themselves are not rigid rules that cannot be violated.

1. Avoid abrupt transitions in element size and geometry. Limit the change in *element stiffness* (approximated by E/V_e, where V_e is the volume of the element) from one element to the next to roughly a factor of 3.
2. Avoid unnecessary element irregularity. Keep aspect ratios (the length ratio of the longest side to the shortest side) below 10:1. Interior angles of quadrilaterals should be as regular as possible. They should not exceed 150° and they should not be less than 30°. Midside nodes on quadratic elements should be within the middle third of the edge.
3. Maintain compatibility between elements. For instance, it is not appropriate to attach one quadratic quadrilateral to two linear

quadrilaterals simply because they have three nodes in common. Such an assembly would not maintain compatibility because of the difference in displacement interpolation on the two sides of the boundary (see Fig. 4.18).

4. Use a fine mesh in regions of high stress gradient (stress concentration). Use a coarse mesh where gradients are low.

5. When Choleski decomposition, or any other *band* solver, is used, minimize the bandwidth of the assembled structure stiffness matrix by proper node numbering. The nonzero entries in the structure stiffness matrix are clustered about the diagonal in a *band*. The *bandwidth* is the number of terms across a row (or down a column) of the band. The *half-bandwidth* is the number of terms from the diagonal out to the edge of the band. The nodal half-bandwidth is computed as $(n_{max} - n_{min} + 1)$, where n_{max} and n_{min} are the largest and smallest structure node numbers in the incidence list for an element. Hence, to minimize bandwidth, keep the range of node numbers that define the incidences for a single element as small as possible. Examples of poor and good node numbering schemes are illustrated in Fig. 4.19.

6. Exploit symmetry in the geometry and loads of the physical system to build the smallest reasonable model.

The finite element method and its use in engineering practice are evolving continuously. For instance, not long ago, material and/or geometric nonlinear analyses were rarely attempted. Today, such analyses are not limited to research but are performed by practicing engineers as well. The popularity of the finite element method is due primarily to the greater availability, and affordability, of user-friendly software that integrates sophisticated analysis capabilities with solid modeling and CAD. Unfortunately, user training and experience is not always

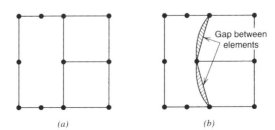

Figure 4.18 Assembly of incompatible elements. (*a*) Undistorted assembly. (*b*) Loss of compatibility under distortion.

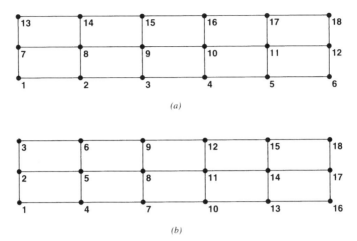

(a)

(b)

Figure 4.19   Node numbering to minimize bandwidth. (a) Poor numbering scheme, half-band-width = 8. (b) Good numbering scheme, half-bandwidth = 5.

equal to the capabilities of the software. Hence, the danger exists that these powerful analytical tools will be used as *black boxes,* without proper understanding of the physical system or of the algorithms used in the analysis. There is no substitute for common sense and sound judgment, and one should remain skeptical of computer-generated results until they can be verified by other means.

An effective means for an engineer to gain experience in performing finite element analysis and to develop confidence in the finite element program is to solve a series of relatively simple *benchmark* problems. Such problems are specially designed to test the accuracy of the individual elements in the program. However, they can also be used as a training device for novice users. A reasonable set of benchmark problems has been proposed by MacNeal and Harder (1984, 1985). Additional problems can be found in (AIAA, 1985).

PROBLEMS

Section 4.2

4.1   A transverse load P is applied to the end of a cantilever beam (Fig. P4.1). The beam has length L, moment of inertia I, and modulus of elasticity E. The displaced shape of the beam is assumed to be of the following forms:

Figure P4.1

(1) $v(x) = a_0 + a_1 x + a_2 x^2$

(2) $v(x) = b_0 \left(1 - \cos \dfrac{\pi x}{2l} \right)$

(3) $v(x) = c_0 + c_1 x + c_2 x^2 + c_3 x^3$

Consider only strain energy due to bending as given by Eq. (4.65) and the potential of the load $[\Omega = -Pv(L)]$ with respect to the undeformed beam.

(a) To the extent possible, simplify each of the assumed displaced shapes to account for the boundary conditions.

(b) Calculate the elastic strain energy U and the potential Ω of the external load P for each of the assumed displaced shapes.

(c) Solve for the parameters $(a_0, \ldots, c_3)$ using the principle of stationary potential energy, where for equilibrium $\delta\Pi = \delta U + \delta\Omega = 0$. *Hint:* The virtual displacement δv is first written in terms of a variation in the parameters $(\delta a_0, \ldots, \delta c_3)$. Then simultaneous equations are written from

$$\delta\Pi = \frac{\partial\Pi}{\partial a_0}\, \partial a_0 + \cdots = 0$$

(d) Compute values of Π and $v(L)$ for each of the assumed displaced shapes. Compare the values of $v(L)$ to each other and to the elasticity solution of $v(L) = PL^3/3EI$.

(e) Discuss the results.

4.2 For the constant-strain triangle element shown in Fig. P4.2:

(a) Write the shape function for each node.

(b) Evaluate each shape function at point P.

(c) Show, numerically for each shape function, that the value of the shape function for node i is equal to the ratio A_{Pjk}/A_{ijk}, where A_{Pjk} is the area of triangle Pjk and A_{ijk} is the area of the element.

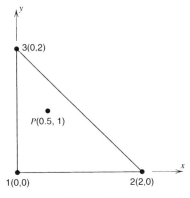

Figure P4.2

4.3 For the mesh shown in Fig. 4.5, construct the Boolean connectivity matrix $[M]$ for elements 2, 3, and 4. Refer to Example 4.2.

4.4 For the mesh shown in Fig. 4.5, assemble the complete stiffness matrix for the structure. Use the notation k_{ij}^e to represent each stiffness coefficient, where the superscript identifies the element number. Refer to Example 4.2.

Section 4.4

4.5 A four-node isoparametric element has nodes at the following (x, y) coordinates: 1(0,0), 2(1,0), 3(2,2), 4(0,1).

  (a) Sketch to scale the element and the lines for which $\xi = \pm\frac{1}{2}$, $\xi = \pm\frac{1}{4}$, $\eta = \pm\frac{1}{2}$, and $\eta = \pm\frac{1}{4}$.

  (b) Write the coordinate interpolation functions: $x(\xi, \eta) = \sum_{i=1}^{4} N_i(\xi, \eta)x_i$ and $y(\xi, \eta) = \sum_{i=1}^{4} N_i(\xi, \eta)y_i$.

  (c) Compute the terms in the Jacobian matrix $[J]$ given by Eq. (4.53).

  (d) Evaluate $|J|$ at $\xi = 0$, $\eta = 0$. Compare this value to the ratio of the area of the element in (x, y) coordinates to that in (ξ, η) coordinates.

4.6 For the linear isoparametric element shown in Fig. P4.6, compute $[B_1]$ at the point $\xi = 0$, $\eta = 0$.

4.7 Using the one-, two-, and three-point Gauss quadrature rules, evaluate the following integrals numerically. Compare the numerical results to the exact solutions.

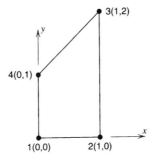

Figure P4.6

(a) $I = \int_{-1}^{1} (6x^3 - 4x^2 + 3x - 2) \, dx$

(b) $I = \int_{-1}^{1} \cosh \xi \, d\xi$

(c) $I = \int_{-1}^{1} e^{\xi} \, d\xi$

4.8 Using the one-, two-, and three-point symmetric Gauss quadrature rules, evaluate the following integrals numerically. Compare the numerical results to the exact solutions.

(a) $I = \int_{-1}^{1} \int_{-1}^{1} \cos \xi \cos \eta \, d\xi \, d\eta$

(b) $I = \int_{-1}^{1} \int_{-1}^{1} \sin^2 \xi \cos \eta \, d\xi \, d\eta$

Section 4.5

4.9 Derive the equivalent nodal load vector for an axial rod element subjected to a concentrated axial force $\bar{P}_C$ acting at L_c from node 1 (see Fig. P4.9).

4.10 Derive the equivalent nodal load vector for an axial rod subjected to a uniformly distributed axial force of magnitude $\bar{q}_0$ acting over the domain $L_a < \bar{x} < L_b$ (see Fig. P4.10).

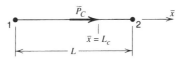

Figure P4.9

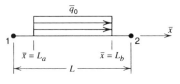

Figure P4.10

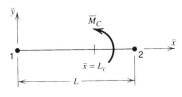

Figure P4.11

4.11 Derive the equivalent nodal load vector for a beam element subjected to a concentrated bending moment $\overline{M}_C$ acting at L_c from node 1 (see Fig. P4.11).

REFERENCES

AIAA. 1985. *Proc. Finite Element Standards Forum,* AIAA/ASME/ASCE/ AHS 26th Structural Dynamics and Materials Conference, Apr. 15, Orlando, FL.

Boresi, A. P., and Chong, K. P. 2000. *Elasticity in Engineering Mechanics,* 2nd ed., Wiley, New York.

Boresi, A. P., Schmidt, R. J., and Sidebottom, O. M. 1993. *Advanced Mechanics of Materials,* Wiley, New York.

Clough, R. W. 1960. The finite element method in plane stress analysis, *Proc. 2nd ASCE Conference on Electronic Computation,* Pittsburgh, PA, pp. 345–378.

Cook, R. D., Malkus, D. S., and Plesha, M. E. 1989. *Concepts and Applications of Finite Element Analysis,* 3rd ed., Wiley, New York.

Courant, R. 1943. Variational methods for the solution of problems of equilibrium and vibrations, *Bull. Am. Math. Soc.,* **49**, 1–23.

Courant, R. 1950. *Differential and Integral Calculus,* Wiley, New York.

MacNeil, R. H., and Harder, R. L. 1984. A proposed standard set of problems to test finite element accuracy, *Proc. 25th Structural Dynamics and Ma-*

terials Conference, AIAA/ASME/ASCE/AHS, May 14, Palm Springs, CA.

MacNeil, R. H., and Harder, R. L. 1985. A proposed standard set of problems to test finite element accuracy, *Finite Elements Anal. Des.,* **1**(1).

Turner, M. J., Clough, R. W., Martin, H. C., and Topp, L. J. 1956. Stiffness and deflection analysis of complex structures, *J. Aeronaut. Sci.* **25**(9), 805–823.

Zienkiewicz, O. C., and Taylor, R. L. 1989. *The Finite Element Method,* 4th ed., McGraw-Hill, New York.

BIBLIOGRAPHY

Baker, A. J., and Pepper, D. W. 1991. *Finite Elements 1–2–3,* McGraw-Hill, New York.

Bathe, K.-J. 1995. *Finite Element Procedures,* Prentice Hall, Upper saddle River, NJ.

Belytschko, T., Liu, W. K., and Moran, B. 2000. *Nonlinear Finite Elements for Continua and Structures,* Wiley, New York.

Bickford, W. B. 1990. *A First Course in the Finite Element Method,* Richard D. Irwin, Homewood, IL.

Burnett, D. S. 1987. *Finite Element Analysis, from Concepts to Applications,* Prentice Hall, Upper Saddle River, NJ.

Cheung, Y. K., Lo, S. H., and Leung, A. Y. T. 1996. *Finite Element Implementation,* Blackwell Science, Malden, MA.

Cook, R. D., Malkus, D. S., and Plesha, M. E. 1989. *Concepts and Applications of Finite Element Analysis,* 3rd ed., Wiley, New York.

Crisfield, M. A. 1996. *Nonlinear Finite Element Analysis of Solids and Structures,* Wiley, New York.

Finlayson, B. A. 1992. *Numerical Methods for Problems with Moving Fronts,* Ravenna Park Publishing, Seattle, WA.

Ghali, A., and Neville, A. M. 1990. *Structural Analysis: A Unified Classical and Matrix Approach,* 3rd ed., Chapman & Hall, London.

Grandin, H., Jr. 1991. *Fundamentals of the Finite Element Method,* Waveland Press, Prospect Heights, IL.

Huebner, K. H. 1994. *The Finite Element Method for Engineers,* 3rd ed., Wiley, New York.

Hughes, T. J. R. 2000. *The Finite Element Method: Linear Static and Dynamic Finite Element Analysis,* Dover Publications, Mineola, NY.

Melosh, R. J. 1990. *Structural Engineering Analysis by Finite Elements,* Prentice Hall, Upper Saddle River, NJ.

Potts, J. F., and Oler, J. W. 1989. *Finite Element Applications with Micro-computers,* Prentice Hall, Upper Saddle River, NJ.

Przemieniecki, J. S. 1985. *Theory of Matrix Structural Analysis,* Dover Publications, Mineola, NY.

Rao, S. S. 1989. *The Finite Element Method in Engineering,* 2nd ed., Pergamon Press, Oxford.

Reddy, J. N. 1993. *An Introduction to the Finite Element Method,* 2nd ed., McGraw-Hill, New York.

Sack, R. L. 1994. *Matrix Structural Analysis,* Waveland Press, Prospect Heights, IL.

Segerland, L. J. 1984. *Applied Finite Element Analysis,* Wiley, New York.

Stasa, F. L., 1995. *Applied Finite Element Analysis for Engineers,* HBJ College & School Division, Holt, Rinehart and Winston, Austin, TX.

Weaver, W., Jr., and Gere, J. M. 1990. *Matrix Analysis of Framed Structures,* 3rd rd., Van Nostrand Reinhold, New York.

Yang, T. Y. 1986. *Finite Element Structural Analysis,* Prentice Hall, Upper Saddle River, NJ.

Zienkiewicz, O. C., and Taylor, R. L. 1989. *The Finite Element Method,* 4th ed., McGraw-Hill, New York.

5

Specialized Methods

5.1 INTRODUCTION

The finite strip method, pioneered in 1968 by Y. K. Cheung (1968a,b), is an efficient tool for analyzing structures with regular geometric platform and simple boundary conditions. Basically, the finite strip method reduces a two-dimensional problem to a one-dimensional problem. In some cases computational savings by a factor of 10 or more are possible compared to the finite element method (Cheung and Tham, 1998).

Originally, the finite strip method was designed for rectangular plate problems [similar to Levy's solution; Timoshenko and Woinowsky-Krieger (1971)]. Later, the finite strip method was extended to treat curved plates (Cheung, 1969b), skewed (quadrilateral) plates, folded plates, and box girders. Formulated as an eigenvalue problem, the finite strip method can be applied to vibration and stability problems of plates and shells with relative ease. Finite prism and finite layer methods were also introduced by Cheung and Tham (1998). These methods reduce three-dimensional problems to two- and one-dimensional problems, respectively, by choosing an appropriate choice of displacement functions. In addition, composite structures, such as sandwich panels with cold-formed facings, can be analyzed efficiently by coupling the finite strip method with the finite prism or finite layer methods (Cheung et al., 1982b; Chong, 1986; Chong et al., 1982a,b; Tham et al., 1982).

The finite strip method has been modified through the use of spline functions to analyze plates of arbitrary shape (Cheung et al., 1982a;

Chong and Chen, 1986; Li et al., 1986; Tham et al., 1986; Yang and Chong, 1982, 1984). Complicated boundary conditions can be accommodated. In general, finite strip methods based on spline functions are more involved than the finite strip method based on trigonometric series. However, they are still more efficient than finite element methods, and they require less input and less computational effort.

5.2 FINITE STRIP METHOD

The finite strip method (FSM), first proposed by Y. K. Cheung (1968a,b), is approaching a state of maturity as a structural analysis technique (Puckett et al., 1987; and Wiseman et al., 1987). Two comprehensive books on the method and its applications are available (Cheung and Tham, 1998; Loo and Cusens, 1978), as are papers on advances in the field (Cheung, 1981; and others). The paper by Wiseman et al., (1987), summarizes developments in the methods of finite strips, finite layers, and finite prisms, and includes a review of 114 references.

To examine the finite strip method, consider a rectangular plate with x and y axes in the plane of the plate and axis z in the thickness direction (Fig. 5.1). Let the corresponding displacement components be denoted by (u, v, w). Then similar to Levy's solution (Timoshenko and Woinowsky-Krieger, 1971), for a typical strip (Fig. 5.1), the w displacement component is (Cheung and Tham, 1998).

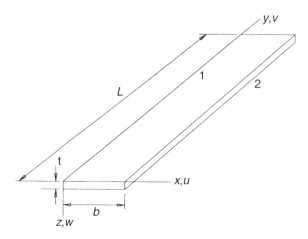

Figure 5.1 Rectangular bending strips.

$$w(x, y) = \sum_{m-1}^{r} f_m(x)Y_m(y) \tag{5.1}$$

in which the functions $f_m(x)$ are polynomials and the functions $Y_m(y)$ are trigonometric terms that satisfy the end conditions in the y direction. The functions Y_m can be taken as basic functions (mode shapes) of the beam vibration equation

$$\frac{d^4Y}{dx^4} - \frac{\mu^4}{a^4} Y = 0 \tag{5.2}$$

where a is the beam (strip) length and μ is a parameter related to frequency, material, and geometric properties.

The general solution of Eq. (5.2) is

$$Y(y) = C_1 \sin \frac{\mu y}{a} + C_2 \cos \frac{\mu y}{a} + C_3 \sinh \frac{\mu y}{a} + C_4 \cosh \frac{\mu y}{a} \tag{5.3}$$

Four boundary conditions are needed to determine the coefficients C_1 to C_4. For example, for both ends simply supported,

$$Y(0) = Y''(0) = Y(a) = Y''(a) = 0 \tag{5.4}$$

Equations (5.3) and (5.4) yield the mode shape functions

$$Y_m(y) = \sin \frac{\mu_m y}{a} \qquad m = 1, 2, 3, \ldots, r \tag{5.5}$$

where $\mu_m = m\pi$; $m = 1, 2, 3, \ldots, r$.

Since the functions Y_m are mode shapes, they are orthogonal (Meirovitch, 1986); that is, they satisfy the relations

$$\int_0^a Y_m Y_n \, dy = 0 \qquad \text{for } m \neq n \tag{5.6}$$

Also, it can be shown that (Cheung and Tham, 1998),

$$\int_0^a Y''_m Y''_n \, dy = 0 \qquad \text{for } m \neq n \tag{5.7}$$

The orthogonal properties of Y_m result in structural matrices with very narrow bandwidths, thus minimizing computational storage and computational time.

Similarly to the finite element method, the functions $f_m(x)$ in Eq. (5.1) can be expressed as

$$f_m(x) = [[C_1] \quad [C_2]] \left\{ \begin{array}{c} \{\delta_1\} \\ \{\delta_2\} \end{array} \right\}_m \tag{5.8}$$

where subscripts 1 and 2 denote sides 1 and 2 of the plate (strip), respectively; $[C_1]$ and $[C_2]$ are interpolating functions, equivalent to shape functions in one-dimensional finite elements; and $\{\delta_1\}$ and $\{\delta_2\}$ are nodal parameters.

Let b be the width of a strip in the plate and let $\bar{x} = x/b$. Taking the functions $f_m(x)$ as linear functions of x, and considering nodal displacements only, we have

$$\delta_1 = w_1, \qquad \delta_2 = w_2 \qquad C_1 = 1 - \bar{x} \qquad C_2 = \bar{x}$$

Hence, by Eq. (5.8), we have

$$f_m(x) = [(1 - \bar{x}) \quad \bar{x}] \left\{ \begin{array}{c} w_1 \\ w_2 \end{array} \right\} = (1 - \bar{x})w_1 + \bar{x}w_2 \tag{5.9}$$

Equation (5.9) is equivalent to a one-dimensional linear finite element (Chapter 4), which employs nodal displacements only.

For higher-order functions, with nodal displacements w_i and first derivatives (nodal slopes) $\theta_i = w_i'$, we have

$$\{\delta_1\} = \left\{ \begin{array}{c} w_1 \\ \theta_1 \end{array} \right\} \qquad \{\delta_2\} = \left\{ \begin{array}{c} w_2 \\ \theta_2 \end{array} \right\}$$

$$[C_1] = [(1 - 3\bar{x}^2 + 2\bar{x}^3) \quad x(1 - 2\bar{x} + \bar{x}^2)] \tag{5.10}$$

$$[C_2] = [(3\bar{x}^2 - 2\bar{x}^3) \quad x(\bar{x}^2 - \bar{x})]$$

By Eqs. (5.8) and (5.10), we obtain

$$f_m(x) = [(1 - 3\bar{x}^2 + 2\bar{x}^3) \quad x(1 - 2\bar{x} + \bar{x}^2) \quad (3\bar{x}^2 - 2\bar{x}^3) \quad x(\bar{x}^2 - \bar{x})]$$

$$\times \left\{ \begin{array}{c} w_1 \\ \theta_1 \\ w_2 \\ \theta_2 \end{array} \right\} \tag{5.11}$$

Equation (5.11) is equivalent to a one-dimensional beam element, which employs both nodal displacements and nodal slopes. Other higher-order functions can be derived in a similar manner (Cheung and Tham, 1998).

5.3   FORMULATION OF THE FINITE STRIP METHOD

In general [see Eqs. (5.1) and (5.8)], the displacement function may be written as

$$w = \sum_{m=1}^{r} Y_m \sum_{k=1}^{s} [C_k]\{\delta_k\}_m \tag{5.12}$$

in which r is the number of mode shape functions [Eq. (5.5)] and s is the number of nodal line parameters.

Let

$$[N_k]_m = Y_m[C_k] \tag{5.13}$$

Then, by Eq. (5.12),

$$w = \sum_{m=1}^{r} \sum_{k=1}^{s} [N_k]_m\{\delta_k\}_m = [N]\{\delta\} \tag{5.14}$$

where $[N]$ denotes the shape functions and $\{\delta\}$ denotes the nodal parameters.

The formulation of the finite strip method is similar to that of the finite element method (Section 4.2). For example, for a strip subjected to bending, the strain matrix (vector), $\{\varepsilon\}$, is given by

$$\{\varepsilon\} = \begin{Bmatrix} M_x \\ M_y \\ M_{xy} \end{Bmatrix} = \begin{Bmatrix} -\partial^2 w/\partial x^2 \\ -\partial^2 w/\partial y^2 \\ 2\partial^2 w/\partial x\,\partial y \end{Bmatrix} \tag{5.15}$$

where M_x, M_y, M_{xy} are moments per unit length.

Differentiating w, Eq. (5.14), and substituting into Eq. (5.15), we obtain the result

$$\{\varepsilon\} = [B]\{\delta\} = \sum_{m=1}^{r} [B]_m \{\delta\}_m \tag{5.16}$$

in which

$$[B]_m = \left\{ \begin{array}{c} -\partial^2[N]_m/\partial x^2 \\ -\partial^2[N]_m/\partial y^2 \\ 2\partial^2[N]_m/\partial x\,\partial y \end{array} \right\} \tag{5.17}$$

Equation (5.16) is similar to Eq. (4.25).

By Hooke's law (Boresi and Chong, 2000) and Eq. (5.16), the stress matrix (vector), $\{\sigma\}$, is

$$\{\sigma\} = [D]\{\varepsilon\}$$

$$= [D] \sum_{m=1}^{r} [B]_m \{\delta\}_m \tag{5.18}$$

where $[D]$ is the elasticity matrix, defined in Section 4.2 [Eqs. (4.10), (4.11), and (4.12)] for isotropic plane strain and plane stress problems.

Minimization of the total potential energy gives (Cheung and Tham, 1998),

$$[K]\{\delta\} = \{F\} \tag{5.19}$$

where

$$[K] = \int_{\text{vol}} [B]^{\mathrm{T}}[D][B]\ dv \tag{5.20}$$

is the stiffness matrix and $\{F\}$ is the load vector.

For a distributed load $\{q\}$, the load vector is

$$\{F\} = \int_{A} [N]^{\mathrm{T}}\{q\}\ dA \tag{5.21}$$

Expanding Eq. (5.20), we obtain the stiffness matrix in the form

$$[K] = \int_{vol} [[B]_1^T[B]_2^T \cdots [B]_r^T][D][[B]_1[B]_2 \cdots [B]_r] \, dv$$

$$= \int_{vol} \begin{bmatrix} [B]_1^T[D][B]_1 & [B]_1^T[D][B]_2 \cdots [B]_1^T[D][B]_r \\ \vdots & \vdots & \vdots \\ [B]_r^T[D][B]_1 & [B]_r^T[D][B]_2 \cdots [B]_r^T[D][B]_r \end{bmatrix} dv$$

$$= \begin{bmatrix} [k]_{11} & [k]_{12} \cdots [k]_{1r} \\ \vdots & \vdots & \vdots \\ [k]_{r1} & [k]_{r2} \cdots [k]_{rr} \end{bmatrix} \tag{5.22}$$

where

$$[k]_{mn} = \int_{vol} [B]_m^T[D][B]_n \, dv \tag{5.23}$$

For each strip (with s nodal line parameters),

$$[k]_{mn} = \begin{bmatrix} [k_{11}] & [k_{12}] \cdots [k_{1s}] \\ \vdots & \vdots & \vdots \\ [k_{s1}] & [k_{s2}] \cdots [k_{ss}] \end{bmatrix}_{mn}$$

In general, the individual elements of matrix $[k]_{mn}$ are given by

$$[k_{ij}]_{mn} = \int_{vol} [B_i]_m^T[D][B_j]_n \, dv \tag{5.24}$$

Similarly the basic element of the load vector (for distributed loads) is

$$\{F_i\} = \int_A [N_i]_m^T\{q\} \, dA \tag{5.25}$$

For simple functions Eqs. (5.24) and (5.25) can be evaluated in closed form (Cheung and Tham, 1998). Alternatively, they can be integrated numerically using Gaussian quadrature or other numerical integration methods.

5.4 EXAMPLE OF THE FINITE STRIP METHOD

In this section the finite strip method is applied to the analysis of a uniformly loaded plate simply supported on two parallel edges. Each nodal line is free to move in the z direction and rotate about the y axis (Fig. 5.1). Thus Eqs. (5.1), (5.5), and (5.10) are to be used. The matrix $[B]_m$, Eqs. (5.17) and (5.23), is (Cheung and Tham, 1998).

$$[B]_m = \begin{bmatrix} \dfrac{6}{b^2}(1 - 2\bar{x})Y_m & \dfrac{2}{b}(2 - 3\bar{x})Y_m \\[2mm] -(1 - 3\bar{x}^2 + 2\bar{x}^3)Y''_m & -x(1 - 2\bar{x} + \bar{x}^2)Y''_m \\[2mm] \dfrac{2}{b}(-6\bar{x} + 6\bar{x}^2)Y'_m & 2(1 - 4\bar{x} + 3\bar{x}^2)Y'_m \end{bmatrix}$$
$$\begin{matrix} \dfrac{6}{b^2}(-1 + 2\bar{x})Y_m & \dfrac{2}{b}(-3\bar{x} + 1)Y_m \\[2mm] -(3\bar{x}^2 - 2\bar{x}^3)Y''_m & -x(\bar{x}^2 - \bar{x})Y''_m \\[2mm] \dfrac{2}{b}(6\bar{x} - 6\bar{x}^2)Y'_m & 2(3\bar{x}^2 - 2\bar{x})Y'_m \end{matrix}$$

For orthotopic plates, the elasticity matrix is (Boresi and Chong, 2000; Cheung and Tham, 1998)

$$[D] = \begin{bmatrix} D_x & D_1 & 0 \\ D_1 & D_y & 0 \\ 0 & 0 & D_{xy} \end{bmatrix} \tag{5.26}$$

where

$$D_x = \frac{E_x t^3}{12(1 - \nu_x \nu_y)} \qquad D_y = \frac{E_y t^3}{12(1 - \nu_x \nu_y)} \qquad D_{xy} = \frac{G t^3}{12}$$

$$D_1 = \frac{\nu_x E_y t^3}{12(1 - \nu_x \nu_y)} = \frac{\nu_y E_x t^3}{12(1 - \nu_x \nu_y)} \tag{5.27}$$

For isotropic materials,

$$E_x = E_y = E, \quad \nu_x = \nu_y = \nu \quad \text{and} \quad G = \frac{E}{2(1 + \nu)} \quad (5.28)$$

By the orthogonal properties, $[k]_{mn} = 0$ if $m \neq n$. Thus, Eq. (5.19) reduces to

$$\begin{bmatrix} [k]_{11} & & & \\ & [k]_{22} & & \\ & & \ddots & \\ & & & [k]_{rr} \end{bmatrix} \begin{Bmatrix} w_1 \\ \theta_1 \\ w_2 \\ \theta_2 \\ \vdots \\ w_{r+1} \\ \theta_{r+1} \end{Bmatrix} = \begin{Bmatrix} \{F_1\} \\ \{F_2\} \\ \vdots \\ \{F_r\} \end{Bmatrix} \quad (5.29)$$

where the matrices $[k]_{mm}$ are 4×4 symmetrical matrices (Cheung and Tham, 1998). The load vector $\{F_m\}$ can be computed by integrating Eq. (5.25). It is a 4×1 matrix given by

$$\{F\} = q \begin{Bmatrix} b/2 \\ b^2/12 \\ b/2 \\ -b^2/12 \end{Bmatrix} \int_0^a \sin \frac{m\pi y}{a} \, dy \quad (5.30)$$

The load matrices $\{F\}_m$ for eccentric and concentric concentrated loads have been derived by Cheung and Tham (1998). With two degrees of freedom (w, θ) along the nodal line, the moments [Eq. (5.15)] are linear functions of x. Thus, to obtain the maximum moment at a known location (e.g., at the center for uniform load and under load point for concentrated load), a nodal line must pass through this location for good accuracy. In a convergence test applied to a uniformly loaded simply supported square plate, Cheung and Tham, 1998) have shown that with four strips (10 equations), greater accuracy was obtained than that by the finite element method with 25 nodes (75 equations). In addition, considerable saving in computational labor is achieved (Yang and Chong, 1984). Higher-order strips have also been applied to plates (Cheung and Tham, 1998). Finite strip formulations for curved and skewed plates have also been derived (Cheung and Tham, 1998). These techniques are applicable to the analysis of curved bridges, stiffened plates, and so on.

5.5 FINITE LAYER METHOD

By selecting functions satisfying the boundary conditions in two directions, the philosophy of the finite strip method can be extended to layered systems. The resulting method is called the *finite layer method* (FLM). The method was first proposed by Cheung and Chakrabarti (1971). The finite layer method is useful for layered materials, rectangular in planform. To illustrate the method, consider Fig. 5.2. Let $\bar{z} = z/c$. Then the lateral displacement component, w, can be expressed as

$$w = \sum_{m=1}^{r} \sum_{n=1}^{t} [(1 - \bar{z})w_{1mn} + \bar{z}w_{2mn}]X_m(x)Y_n(y) \qquad (5.31)$$

in which, w_{1mn} and w_{2mn} are displacement parameters for side 1 (top) and side 2 (bottom), respectively. The displacement component, w, is assumed to vary linearly in the z direction. The functions X_m and Y_n are taken as terms in a trigonometric series, satisfying the boundary conditions. In this manner a three-dimensional problem is reduced to a one-dimensional problem with considerable saving in computer storage and computational time (Cheung, and Tham, 1998; Cheung et al., 1982b).

By linear theory, the (x, y) displacement components (u, v) are linearly related to the derivatives of w; that is,

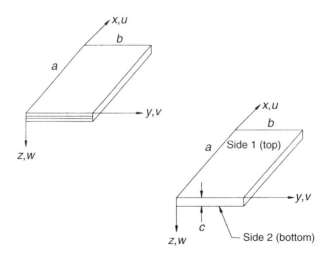

Figure 5.2 Finite layers.

$$u = A \frac{\partial w}{\partial x} \tag{5.32}$$

$$v = B \frac{\partial w}{\partial y} \tag{5.33}$$

Thus, as with Eq. (5.31), we may write

$$v = \sum_{m=1}^{r} \sum_{n=1}^{t} [(1 - \bar{z})v_{1mn} + \bar{z}v_{2mn}]X_m(x)Y_n'(y) \tag{5.34}$$

$$u = \sum_{m=1}^{r} \sum_{n=1}^{t} [(1 - \bar{z})u_{1mn} + \bar{z}u_{2mn}]X_m'(x)Y_n(y)$$

in which u_{1mn}, v_{1mn}, u_{2mn} and v_{2mn} are displacement parameters of sides 1 and 2.

The displacement is

$$\{f\} = \begin{Bmatrix} u \\ v \\ w \end{Bmatrix} = \sum_{m=1}^{r} \sum_{n=1}^{t} [N]_{mn}\{\delta\}_{mn} = [N]\{\delta\} \tag{5.35}$$

where

$$\{\delta\}_{mn} = [u_{1mn}, v_{1mn}, w_{1mn}, u_{2mn}, v_{2mn}, w_{2mn}]^{\mathrm{T}} \tag{5.36}$$

The strain–displacement relationship is (Borcsi and Chong, 2000)

$$\{\varepsilon\} = \begin{Bmatrix} \varepsilon_{xx} \\ \varepsilon_{yy} \\ \varepsilon_{zz} \\ \varepsilon_{xy} \\ \varepsilon_{yz} \\ \varepsilon_{zx} \end{Bmatrix} = \begin{Bmatrix} \partial u/\partial x \\ \partial v/\partial y \\ \partial w/\partial z \\ \frac{1}{2}(\partial u/\partial y + \partial v/\partial x) \\ \frac{1}{2}(\partial v/\partial z + \partial w/\partial y) \\ \frac{1}{2}(\partial w/\partial x + \partial u/\partial z) \end{Bmatrix} = \sum_{m=1}^{r} \sum_{n=1}^{t} [B]_{mn}\{\delta\}_{mn} = [B]\{\delta\}$$

$$\tag{5.37}$$

The stress–strain relationship is

$$\{\sigma\} = [\sigma_{xx} \ \sigma_{yy} \ \sigma_{zz} \ \sigma_{xy} \ \sigma_{yz} \ \sigma_{zx}]^{\mathrm{T}} = [D][B]\{\delta\} \qquad (5.38)$$

Proceeding as in Section 5.4, we obtain the stiffness matrix (Cheung and Tham, 1998)

$$[S] = \int_0^a \int_0^b \int_0^c [B]^{\mathrm{T}}[D][B] \ dx \ dy \ dz \qquad (5.39)$$

For a simply supported rectangular plate, we have the conditions

$$\int_0^a \int_0^b \int_0^c [B]^{\mathrm{T}}_{mn}[D][B]_{rs} \ dx \ dy \ dz = 0 \qquad \text{for } mn \neq rs \quad (5.40)$$

Hence, the off-diagonal terms are zero, and the stiffness matrix reduces to

$$[K] = \int_0^a \int_0^b \int_0^c \begin{bmatrix} [B]^{\mathrm{T}}_{11}[D][B]_{11} \\ [B]^{\mathrm{T}}_{12}[D][B]_{12} \\ \cdot\,\cdot\,\cdot \\ [B]^{\mathrm{T}}_{1t}[D][B]_{1t} \\ [B]^{\mathrm{T}}_{21}[D][B]_{21} \\ \cdot\,\cdot\,\cdot \\ [B]^{\mathrm{T}}_{rt}[D][B]_{rt} \end{bmatrix} dx \ dy \ dz \qquad (5.41)$$

The remaining formulation is similar to that of Sections 5.3 and 5.4.

5.6 FINITE PRISM METHOD

Similarly to Eq. (5.1), for prismatic members with rectangular plan-form, the three-dimensional problem can be reduced to a two-dimensional problem by expressing the displacement function f as (Cheung and Tham, 1998)

$$f = \sum_{m=1}^{r} f_m(x, z)Y_m \qquad (5.42)$$

where $f_m(x, z)$ is a function of (x, z) only and Y_m are trigonometric functions of y, which satisfy the boundary conditions in the y direction. In formulating the finite prism method (FPM), including nodal displacements and shape functions as in the finite element method, it is convenient to use nodal coordinates. Therefore, let ξ, η be the local coordinates of an element, ξ_k, η_k the nodal coordinates of the element, δ the displacement of a point in the element, δ_k the nodal displacements, and (ϕ_k, ψ_k) functions associated with a particular coordinate system (such as Cartesian, skew, or curvilinear). Then we can represent the local coordinates in terms of nodal coordinates in the form

$$\xi = \sum_{k=1}^{s} \phi_k \xi_k \tag{5.43}$$

where s refers to the number of nodes of the element.

Similarly, the displacement δ of a point within the element can be expressed in terms of the nodal displacements δ_k as

$$\delta = \sum_{k=1}^{s} \psi_k \delta_k \tag{5.44}$$

In general, $\phi_k \neq \psi_k$; however, if $\phi_k = \psi_k$, the element is termed *isoparametric*. Using Eq. (5.42), for a prismatic member with two ends simply supported, the lateral (out-of-plane) displacement component is

$$w = \sum_{m=1}^{r} w_m(x, z) \sin k_m y \tag{5.45}$$

in which

$$w_m(x, z) = \sum_{k=1}^{s} C_k w_{km} \tag{5.46}$$

and w_{km} are the mth term nodal displacements at the kth node. the C_k are the shape functions for the two-dimensional element (in the ξ–η plane). For an isoparametric six-node (ISW'6) model (Fig. 5.3), the shape functions are given as follows:

Corner nodes:

$$C_k = \tfrac{1}{4}\eta_k\eta(1 + \eta_k\eta)(1 + \xi_k\xi) \tag{5.47}$$

Midside nodes:

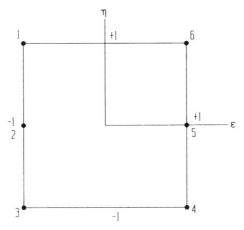

Figure 5.3 ISW'6 model.

$$C_k = \tfrac{1}{2}(1 + \xi_k \xi)(1 - \eta^2) \tag{5.48}$$

Therefore, the lateral (out-of-plane) displacement component for the isoparametric six-node model is

$$w = \sum_{m=\mathrm{i}}^{r} \sum_{k=1}^{6} C_k w_{km} \sin k_m y \tag{5.49}$$

The corresponding in-plane displacement components (u, v) are

$$u = \sum_{m=1}^{r} \sum_{k=1}^{6} C_k u_{km} \sin k_m y \tag{5.50}$$

$$v = \sum_{m=1}^{r} \sum_{k=1}^{6} C_k v_{km} \cos k_m y \tag{5.51}$$

The stiffness matrix is developed in a manner similar to the development in Sections 5.3 and 5.4. It is (Chong et al., 1982b)

$$^{\mathrm{P}}K_{ijmn} = \int {}^{\mathrm{P}}B_{im}^{\mathrm{T}} \, {}^{\mathrm{P}}D \, {}^{\mathrm{P}}B_{jn} \, d(\mathrm{vol}) \tag{5.52}$$

where

$$
{}^{P}B_{im} = \left\{
\begin{array}{ccc}
\dfrac{\partial C_i}{\partial x} \sin \dfrac{m\pi y}{L} & 0 & 0 \\[2ex]
0 & -C_i \dfrac{m\pi}{L} \sin \dfrac{m\pi y}{L} & 0 \\[2ex]
0 & 0 & \dfrac{\partial C_i}{\partial z} \sin \dfrac{m\pi y}{L} \\[2ex]
C_i \dfrac{m\pi}{L} \cos \dfrac{m\pi y}{L} & \dfrac{\partial C_i}{\partial x} \cos \dfrac{m\pi y}{L} & 0 \\[2ex]
0 & \dfrac{\partial C_i}{\partial x} \cos \dfrac{m\pi y}{L} & C_i \dfrac{m\pi}{L} \cos \dfrac{m\pi y}{L} \\[2ex]
\dfrac{\partial C_i}{\partial x} \sin \dfrac{m\pi y}{L} & 0 & \dfrac{\partial C_i}{\partial z} \sin \dfrac{m\pi y}{L}
\end{array}
\right\}
$$

$$(5.53)$$

${}^{P}D$ is the elasticity matrix for isotropic materials. In Eq. (5.52), the superscript P indicates the finite prism model.

The geometric transformation from the natural coordinate (x, z) to the local coordinate (ξ, η) can be carried out as in the standard finite element method, and the stiffness matrix can be obtained accordingly (Cheung and Tham, 1998).

5.7 APPLICATIONS AND DEVELOPMENTS OF FSM, FLM, AND FPM

Owing to their narrow bandwidth and reduction in dimensions, the finite strip, finite layer, and finite prism methods (FSM, FLM, and FPM) are especially adaptable for personal computers (Rhodes, 1987) and minicomputers, with significant savings in computational labor. Developments and applications are presented by Wiseman et al. (1987) and Graves-Smith (1987).

One of the first reported applications of the method was to orthotropic right box girder bridge decks (Cheung, 1969a). Cheung suggested that the finite strip method could be used in a composite slab-beam system, where the longitudinal beam stiffness could be derived separately and added to the structure stiffness. This procedure is a refinement to systems in which the action of longitudinal beams is approximated by assuming orthotropic properties for the slab. Cheung further asserted that the beam would have to coincide with a strip nodal line. This condition was verified by comparison with theoretical and model test results for simply supported slab-beam models with varying slab thicknesses (Cheung et al., 1970). A force method for introducing rigid column supports in the strips was also described.

An analytical method using a separation-of-variables procedure and series solution for the resulting differential equations was developed by Harik and his associates (1984, 1985, 1986) for orthotropic sectorial plates and for rectangular plates under various loadings. The plates were divided into strips where point and patch loads were applied, and the accuracy of the solutions was demonstrated for various boundary conditions.

Two refinements aimed at increasing the accuracy of finite strips in the in-plane transverse direction were developed by Loo and Cusens (1970). One refinement ensued curvature compatibility at the nodal lines. Although this requirement limited the analysis to plates with uniform properties across the section, convergence was shown to be faster for appropriate plate systems. The other refinement introduced an auxiliary nodal line within the strips. A drawback of these refinements is that they both increase the size of the stiffness matrix by 50%.

Other formulations for increasing the accuracy or the range of applicability of the finite strip method have been published. Brown and Ghali (1978) used a subparametric finite strip for the analysis of quadrilateral plates. Bucco et al. (1979) proposed a deflection contour method which allows the finite strip method to be used for any plate shape and loading for which the deflection contour is known.

Other applications of the finite strip method have used spline functions for the boundary component of the displacement function. Yang and Chong (1982, 1984) used X-spline functions, allowing extension of the finite strip method to plate bending problems with irregular boundaries. The solution was shown to converge correctly for a trapezoidal plate, and techniques for approximating a plate with more irregularly shaped sides and ends were discussed. The buckling of irregular plates has been investigated by Chong and Chen (1986). Extensive development of the finite strip method using the cubic B-spline as the boundary component of the displacement function was done by Cheung et al. (1982a). After initial application to flat plates and box girder bridges by Cheung and Fan (1983), the cubic B-spline method was extended to skewed plates by Tham et al (1986), to curved slabs by Cheung et al. (1986) and to arbitrarily shaped general plates by Li et al. (1986). Shallow shells were analyzed by Fan and Cheung (1983) using a spline finite strip formulation based on the theory of Vlasov (1961).

The application of the finite strip method to folded plates by Cheung (1969c) extended the method to three-dimensional plate structures. The strips used were simply supported with respect to out-of-plane displacement and had similar in-plane longitudinal displacement functions. These end conditions suggest a real plate system having end diaphragms or supports with infinite in-plane stiffness. This system is an acceptable approximation for many structures, including roof systems resting on walls and box girder bridges with plate diaphragms.

A study of folded plates, continuous over rigid supports, has been presented by Delcourt and Cheung (1978). They represented longitudinal displacement functions as eigenfunctions of a continuous beam with the same span and relative span stiffnesses as the plate structure. For comparison, the longitudinal displacement function for a structure with pinned ends and no internal supports is a sine function.

The finite strip method has become popular for the analysis of slab girder and box girder bridges. Cheung and Chan (1978) used folded plate elements to study 392 theoretical bridge models in order to suggest refinements to American and Canadian design codes. Cusens and Loo (1974) have done extensive work in the application of the finite strip method to the analysis of prestressed concrete box girder bridges, and their book (Loo and Cusens, 1978) contains an extensive treatment of the subject with many examples.

The analysis of cylindrically orthotropic curved slabs using the finite strip method was described by Cheung (1969c). The primary difference between this technique and the finite strip method for right bridge decks is the use of a polar coordinate system for the displacement function. It was noted that a model in this coordinate system could be used to approximate a rectangular slab by specifying a very large radius for the plate. A radial shell element was subsequently developed (Cheung and Cheung, 1971a) which allowed the analysis of curved box girder bridges supported by rigid end diaphragms.

Because of its efficiency and ease of input, the finite strip method has been readily adapted to free vibration and buckling problems. The formulation for frequency analysis was outlined by Cheung et al. (1971). Characteristic functions for strips with varying end conditions have also been developed (Cheung and Cheung, 1971b). Finite strips and prisms were combined by Chong et al. (1982a,b) to study free vibrations and to calculate the effects of different face temperatures on foam-core architectural sandwich panels (Fig. 5.4). The panel facings were modeled by finite strips and the panel core by finite prisms. The eigenvalue problem for free vibrations is characterized by the equation (Chong et al., 1982a)

$$-\overline{\omega}^2 \, {}^{S}M_m^P \, {}^{S}\overline{\delta}_m^P - \overline{\omega}^2 \, {}^{S}M_m^B \, {}^{S}\overline{\delta}_m^B - \overline{\omega}^2 \, {}^{P}M_m \, {}^{P}\overline{\delta}_m$$
$$+ \, {}^{S}K_m^P \, {}^{S}\overline{\delta}_m^P + \, {}^{S}K_m^B \, {}^{S}\overline{\delta}_m^B + \, {}^{P}K_m \, {}^{P}\overline{\delta}_m = 0 \quad (5.54)$$

where $\overline{\omega}$ is the frequency, ${}^{S}M_m^P$ is the in-plane mass matrix of the bending strip, ${}^{S}M_m^B$ is the bending mass matrix of the bending strip, ${}^{P}M_m$ is the mass matrix of the finite prism, ${}^{S}K_m^P$ is the in-plane stiffness matrix of the bending strip, ${}^{S}K_m^B$ is the bending stiffness matrix of the bending strip, ${}^{P}K_m$ is the stiffness matrix of the finite prism, ${}^{S}\overline{\delta}_m^P$ and ${}^{S}\overline{\delta}_m^B$ are the in-plane and bending interpolation parameters of the bending strip respectively, and ${}^{P}\overline{\delta}_m$ are the interpolating parameters of the finite prism.

Yoshida (1971) was among the first to develop the geometric stiffness matrix for buckling of rectangular plates. Przemieniecki (1972) used the same displacement function as in the finite strip method (Cheung and Tham, 1998) for the analysis of local stability in plates and stiffened panels. Chong and Chen (1986) investigated the buckling of nonrectangular plates by spline finite strips. Cheung et al. (1982b) used the finite layer method to derive buckling loads of sandwich plates. Mahendran and Murray (1986) modified the standard finite strip displacement function to include out-of-phase displacement parameters, allowing shearing loads to be applied to the strips in the buckling analysis.

The finite strip method has been used as the basis for parametric and theoretical studies of column buckling. Both local and overall buckling of H-columns with residual stress under axial load were studied by Yoshida and Maegawa (1978). Buckling of columns with I-sections was

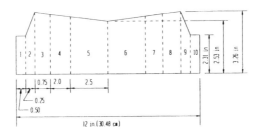

Figure 5.4 Finite strips and prisms in a sandwich panel.

studied by Hancock (1981). The interaction of the two modes of buckling for other prismatic columns was also investigated by Sridharan and Benito (1984) using the finite strip method. Hasegawa and Maeno (1979) determined the effects of stiffeners on cold-formed steel shapes using finite strips. The postbuckling behavior of columns has been the subject of papers by Graves-Smith and Sridharan (1979), Sridharan and Graves-Smith (1981), and Graves-Smith and Gierlinski (1982).

The compound strip method, an enhancement of the flexibility and applicability of the finite strip method, was developed by Puckett and Gutkowski (1986). This formulation added both longitudinal and transverse beams at any location in a strip and included the effect of column supports. A stiffness formulation, which included the added stiffness of all elements associated with a beam, eliminated the repetitive solution passes required for each redundant in a flexibility analysis. The strain energies for flexure and torsion of the beams as well as for axial and flexural deformation of the columns were derived in terms consistent with the strip displacement function, allowing the total strip stiffness to be calculated by adding the element stiffnesses. With the compound strip method, the analysis of continuous beams can be performed using simple displacement functions previously suitable only for single spans. The compound strip method was extended to curved plate systems and was used for dynamic analysis by Puckett and Lang (1986).

Gutkowski et al. (1991) developed a cubic *B*-spline finite strip method for the analysis of thin plates. Spline series with unequal spacing allow local discretization refinement near patch and concentrated loads. Oscillatory convergence (Gibb's phenomenon) is avoided. Spline finite strips have also been used to analyze non-prismatic space structures (Tham, 1990).

Transient responses of noncircular cylindrical shells subject to shock waves were studied by Cheung et al. (1991), using finite strips and a coordinate transformation. Arbitrarily shaped sections were mapped approximately into circular cylindrical shells in which finite strips can be applied for discretization. The finite strip method was used for the free vibration and buckling analysis of plates with abrupt thickness changes and a nonhomogeneous winkler elastic foundation (Cheung et al., 2000).

REFERENCES

Boresi, A. P., and Chong, K. P. 2000. *Elasticity in Engineering Mechanics,* Wiley, New York.

Brown, T. G., and Ghali, A. 1978. Finite strip analysis of quadrilateral plates in bending, *J. Eng. Mech. Div., Proc. ASCE,* **104**(EM2), 481–484.

Bucco, D., Mazumdar, J. and Sved, G. 1979. Application of the finite strip method combined with the deflection contour method to plate bending problems, *Comput. Struct.,* **1,** 827–830.

Cheung, Y. K. 1968a. The finite strip method in the analysis of elastic plates with two opposite simply supported ends, *Proc. Inst. Civ. Eng.,* **40,** 1–7.

Cheung, Y. K. 1968b. Finite strip method analysis of elastic slabs, *J. Eng. Mech. Div., Proc. ASCE,* **94**(EM6), 1365–1378.

Cheung, Y. K. 1969a. Orthotropic right bridges by the finite strip method, Publ. SP-26, American Concrete Institute, Detroit, MI, pp. 182–203.

Cheung, Y. K. 1969b. The analysis of cylindrical orthotropic curved bridge decks, *Proc. Int. Assoc. Bridge Struct. Eng.,* **29,** 41–51.

Cheung, Y. K. 1969c. Folded plate structures by the finite strip method, *J. Struct. Div., Proc. ASCE,* **95**(ST2), 2963–2979.

Cheung, Y. K. 1981. Finite strip method in structural and continuum mechanics, *Int. J. Struct.,* **1,** 19–37.

Cheung, Y. K., and Chakrabarti, S. 1971. Analysis of simply supported thick, layered plates, *J. Eng. Mech. Div., Proc. ASCE,* **97**(EM3), 1039–1044.

Cheung, M. S., and Chan, M. Y. T. 1978. Finite strip evaluation of effective flange width of bridge girders, *Can. J. Civ. Eng.,* **5,** 174–185.

Cheung, M. S., and Cheung, Y. K. 1971a. Analysis of curved box girder bridges by finite strip method, *Proc. Int. Assoc. Bridge Struct. Eng.,* **31**(I), 1–19.

Cheung, M. S., and Cheung, Y. K. 1971b. Natural vibrations of thin, flat-walled structures with different boundary conditions, *J. Sound Vib.,* **18,** 325–337.

Cheung, Y. K., and Fan, S. C. 1983. Static analysis of right box girder bridges by spline finite strip method, *Proc. Inst. Civ. Eng.,* Pt. 2, **75,** 311–323.

Cheung, Y. K., and Tham, L. G. 1998. *The Finite Strip Method,* CRC Press, Boca Raton, FL.

Cheung, M. S., Cheung, Y. K., and Ghali, A. 1970. Analysis of slab and girder bridges by the finite strip method, *Build. Sci.,* **5,** 95–104.

Cheung, M. S., Cheung, Y. K., and Reddy, D. V. 1971. Frequency analysis of certain single and continuous span bridges, in *Developments in Bridge Design and Construction,* Rocky, K. C., Bannister, J. L., and Evans, H. R. (eds.), Crosby Lockwood, London, pp. 188–199.

Cheung, Y. K., Fan, S. C., and Wu, C. Q. 1982a. Spline finite strip in structure analysis, in *Proc. International Conference on Finite Element Methods,* Shanghai, Gordon & Breach, London, pp. 704–709.

Cheung, Y. K., Tham, L. G., and Chong, K. P. 1982b. Buckling of sandwich plate by finite layer method, *Comput. Struct.,* **15**(2), 131–134.

Cheung, Y. K., Tham, L. G., and Li, W. Y. 1986. Application of spline-finite strip method in the analysis of curved slab bridge, *Proc. Inst. Civ. Eng.,* Pt. 2, **81,** 111–124.

Cheung, Y. K., Yuan, C. Z., and Xiong, Z. J. 1991. Transient response of cylindrical shells with arbitrary shaped sections, *J. Thin-Walled Struct.,* **11**(4), 305–318.

Cheung, Y. K., Au, F. T. K., and Zheng, D. Y. 2000. Finite Strip Method for the free vibration and buckling analysis of plates with abrupt changes in thickness and complex support conditions, *Thin-Walled Struct.,* **36,** 89–110.

Chong, K. P. 1986. Sandwich panels with cold-formed thin facings, Keynote paper, *Proc. IABSE International Colloquium on Thin-Walled Metal Structures in Buildings,* Stockholm, Sweden, Vol. 49, pp. 339–348.

Chong, K. P., and Chen, J. L. 1986. Buckling of irregular plates by splined finite strips, *AIAAJ.,* **24**(3), 534–536.

Chong, K. P., Cheung, Y. K., and Tham, L. G. 1982a. Free vibration of formed sandwich panel by finite-prism-strip method, *J. Sound Vib.,* **81**(4), 575–582.

Chong, K. P., Tham, L. G., and Cheung, Y. K. 1982b. Thermal behavior of formed sandwich plate by finite-prism-method, *Comput. Struct.,* **15,** 321–324.

Cusens, A. R., and Loo, Y. C. 1974. Applications of the finite strip method in the analysis of concrete box bridges, *Proc. Inst. Civ. Eng.,* Pt. 2, **57,** 251–273.

Delcourt, C., and Cheung, Y. K. 1978. Finite strip analysis of continuous folded plates, in *Proc. International Association Bridge and Structural Engineers,* May, pp. 1–16.

Fan, S. C., and Cheung, Y. K. 1983. Analysis of shallow shells by spline finite strip method, *Eng. Struct.,* **5,** 255–263.

Graves-Smith, T. R. 1987. The finite strip analysis of structures, in *Developments in Thin-Walled Structures,* Vol. 3, Rhodes, J., and Walker, A. C. (eds.), Elsevier Applied Science, Barking, Essex, England.

Graves-Smith, T. R., and Gierlinski, J. T. 1982. Buckling of stiffened webs by local edge loads, *J. Struct. Div., Proc. ASCE,* **108**(ST6), 1357–1366.

Graves-Smith, T. R., and Sridharan, S. 1979. Elastic collapse of thin-walled columns, in *Recent Technical Advances and Trends in Design, Research, and Construction,* International conference at the University of Strathclyde, Glasgow, Rhodes, J., and Walker, A. C. (eds.), pp. 718–729.

Gutkowski, R. M., Chen, C. J., and Puckett, J. A. 1991. Plate bending analysis by unequally spaced splines, *J. Thin-Walled Struct.,* **11**(4), 413–435.

Hancock, G. J. 1981. Interactive buckling in I-section columns, *J. Struct. Div., Proc. ASCE,* **107,** 165–180.

Harik, I. E. 1984. Analytical solution to orthotropic sector, *J. Eng. Mech.,* **110,** 554–568.

Harik, I. E., and Pashanasangi, S. 1985. Curved bridge desks: analytical strip solution, *J. Struct. Eng.,* **111,** 1517–1532.

Harik, I. E., and Salamoun, G. L. 1986. Analytical strip solution to rectangular plates, *J. Eng. Mech.,* **112,** 105–118.

Hasegawa, A., and Maeno, H. 1979. Design of edge stiffened plates, in *Recent Technical Advances and Trends in Design, Research, and Construction,* International conference at the University of Strathclyde, Glasgow, Rhodes, J., and Walker, A. C. (eds.), pp. 679–680.

Li, W. Y., Cheung, Y. K., and Tham, L. G. 1986. Spline finite strip analysis of general plates, *J. Eng. Mech.,* **112**(1), 43–54.

Loo, Y. C., and Cusens, A. R. 1970. A refined finite strip method for the analysis of orthotropic plates, *Proc. Inst. Civ. Eng.,* **48,** 85–91.

Loo, Y. C., and Cusens, A. R. 1978. *The Finite-Strip Method in Bridge Engineering,* Viewpoint Press, Tehachapi, CA.

Mahendran, M., and Murray, N. W. 1986. Elastic buckling analysis of ideal thin walled structures under combined loading using a finite strip method, *J. Thin-Walled Struct.,* **4,** 329–362.

Meirovitch, L. 1986. *Elements of Vibration Analysis,* 2nd ed., McGraw-Hill, New York.

Przemieniecki, J. S. 1972. Matrix analysis of local instability in plates, stiffened panels and columns, *Int. J. Numer. Methods Eng.,* **5,** 209–216.

Puckett, J. A., and Gutkowski, R. M. 1986. Compound strip method for analysis of plate systems, *J. Struct. Eng.,* **112**(1), 121–138.

Puckett, J. A., and Lang, G. J. 1986. Compound strip method for continuous sector plates, *J. Eng. Mech.,* **112**(5), 1375–1389.

Puckett, J. A., and Lang, G. J. 1989. Compound strip method for the frequency analysis of continuous sector plates, *J. Thin-Walled Struct.,* **8**(3), 165–182.

Puckett, J. A., Wiseman, D. L., and Chong, K. P. 1987. Compound strip method for the buckling analysis of continuous plates, *Int. J. Thin-Walled Struct.,* **5**(5), 385–402.

Rhodes, J. 1987. A simple microcomputer finite strip analysis, in *Dynamics of Structures,* Roesset, J. M. (ed.), American Society of Civil Engineers, Reston, VA, pp. 276–291.

Sridharan, S., and Benito, R. 1984. Columns: static and dynamic interactive buckling, *J. Eng. Mech.,* **110,** 49–65.

Sridharan, S., and Graves-Smith, T. R. 1981. Postbuckling analyses with finite strips, *J. Eng. Mech. Div., Proc. ASCE,* **107**(EM5), 869–888.

Tham, L. G. 1990. Application of spline finite strip method in the analysis of space structures, *J. Thin-Walled Struct.,* **10**(3), 235–246.

Tham, L. G., Chong, K. P., and Cheung, Y. K. 1982. Flexural bending and axial compression of architectural sandwich panels by finite-prism-strip methods, *J. Reinf. Plast. Compos. Mater.,* **1,** 16–28.

Tham, L. G., Li, W. Y., Cheung, Y. K., and Cheng, M. J. 1986. Bending of skew plates by spline-finite-strip method, *Comput. Struct.,* **22**(1), 31–38.

Timoshenko, S. P., and Woinowsky-Krieger, S. 1971. *Theory of Plates and Shells,* 2nd ed., McGraw-Hill, New York.

Vlasov, V. Z. 1961. *Thin-Walled Elastic Beams* (trans. Y. Schechtman, U.S. Department of Commerce, Washington, D.C.

Wiseman, D. L., Puckett, J. A., and Chong, K. P. 1987. Recent developments of the finite strip method, in *Dynamics of Structures,* Roesset, J. M. (ed.), American Society of Civil Engineers, Reston, VA, pp. 292–309.

Yang, H. Y., and Chong, K. P. 1982. On finite strip method, *Proc. International Conference on Finite Element Methods,* Shanghai, China, Department of Civil Engineering, University of Hong Kong, Hong Kong, pp. 824–829.

Yang, H. Y., and Chong, K. P. 1984. Finite strip method with X-spline functions, *Comput. Struct.,* **18**(1), 127–132.

Yoshida, H. 1971. Buckling analysis of plate structures by finite strip method, *Proc. Jpn. Soc. Nav. Arch.,* **130,** 161–171.

Yoshida, H., and Maegawa, K. 1978. Local and member buckling of H-columns, *J. Struct. Mech.,* **6,** 1–27.

BIBLIOGRAPHY

Cheung, Y. K., Lo, S. H., and Leung, A. Y. T. 1996. *Finite Element Implementation,* Blackwell Science, Malden, MA.

6

The Boundary Element Method

6.1 INTRODUCTION

Boundary integral equations, which form the basis of the boundary element method, have been in existence for a long time. The boundary element method as it is known today, however, has been developed largely in the last three decades. The rapid development of the method may have drawn its motivation from the limitations of the finite element method. A major step in performing the finite element analysis is the discretization of the domain into finite elements, called *meshing*. For arbitrarily shaped three-dimensional objects, meshing is an extremely tedious job. Often, this step may take weeks, even months, to accomplish, whereas the rest of the analysis may only require a few days. There are situations where the mesh for the object may need to be defined anew several times for the same analysis. In metal-forming operations, the metal workpiece undergoes very large deformations, including large rotations. The individual elements in the mesh may become severely deformed and possess unduly large aspect ratios. Unless the mesh for the deformed workpiece is redefined several times during the analysis, the original mesh may lead to erroneous results. In the simulation of crack propagation in a solid object, the mesh is first defined with respect to the initial geometry of the crack. If the crack does not propagate along element boundaries, a rare occurrence, the crack will intersect one or more elements. To account for the new location of the crack, the mesh may need to be redefined for each advance of the crack front. In shape optimization problems (Saigal and Kane, 1990), the geometry of the object is revised continually in each calculation step to proceed toward its optimal configuration. For each revision of the geometry, a new mesh needs to be defined. The applications mentioned above and several others make use of the finite element method undesirable. There have been, and continue to be, several attempts in the literature to obviate the problems associated with meshing. Through the use of appropriate mathematical theorems, the boundary element method reduces the dimensions of the problem by one degree. Thus, for a three-dimensional object, a two-dimensional discretization—that of the surfaces bounding the object—is required. Similarly, for two-dimensional analyses, a one-dimensional discretization of the lines enclosing the object is required. This reduction in the dimensions of the problem by one degree leads to significant advantages in terms of ease of discretization of the domain and of reducing the overall time to perform the analysis of objects with complex geometries. Consider the plate with centrally located hole under uniaxial loading shown in Fig. 6.1a. For this problem, the boundary element meshes using quadratic boundary elements are shown in Fig. 6.1b and c, respectively. No boundary elements are shown on the axes of symmetry of the plate in the discretized model shown in Fig. 6.1c. The effects due to symmetry can be included within the analytical formulation of the boundary elements (Saigal et al., 1990a; Kaljevic and Saigal, 1995b). This eliminates the need for providing boundary elements on the axes of symmetry (Kaljevic and Saigal, 1995a). In view of the meshes shown in Fig. 6.1b and c, respectively, the advantage offered by boundary elements in terms of meshing requirements is quite apparent.

There are, however, several limitations of the boundary element method. A true boundary-only formulation is typically obtained only for linear elastic problems and in the absence of body forces. Serious developments, including the particular integral approach (Ahmad and Banerjee, 1986; Henry and Banerjee, 1988a; Henry et al., 1987), the dual reciprocity method (Niku and Brebbia, 1988; Partridge and Brebbia, 1990; Partridge et al., 1991; Wrobel and Brebbia, 1987) and the multiple reciprocity method (Neves and Brebbia, 1991; Nowak and Neves, 1994; Sladek and Sladek, 1996), have been reported that reduce the volume integrals due to body forces into equivalent surface integrals. Similar developments that account for nonlinear effects through boundary-only formulations have also been reported (Henry and Banerjee, 1988b).

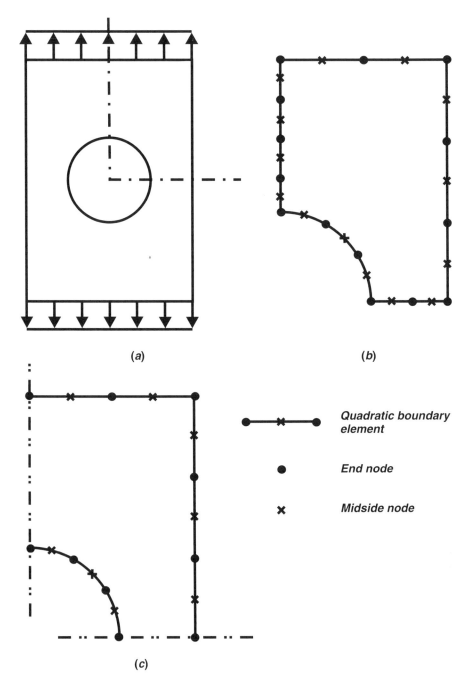

Figure 6.1 Boundary element models of a plate with centrally located hole. (*a*) Plate geometry and loading. (*b*) Boundary element mesh. (*c*) Boundary element mesh for formulations with symmetry conditions included.

A finite element formulation leads to sparse, banded symmetric matrices, for which highly efficient solution procedures are available in the numerical analysis literature. Boundary element formulations, however, lead to dense, fully populated, unsymmetric matrices (Banerjee, 1994). This aspect of boundary element method seriously compromises the advantages due to the reduction of the problem by one degree. There have also been recent developments in the literature toward the development of symmetric boundary element formulations (Hartmann et al., 1985; Maier et al., 1990; Balakrishna et al., 1994; Gray and Paulino, 1997; Layton et al., 1997).

The boundary element method is used today in several industries for the solution of complex analysis problems (Banerjee and Wilson, 1989). It does not, however, appear to have gained the popularity enjoyed by the finite element method. While numerous commercial finite element codes are available to the analyst, only a handful of commercial boundary element codes are in existence. Despite the lack of its popularity, the boundary element method is a powerful tool, and for certain classes of problem the tool of choice for their analysis.

6.2   INTEGRALS IN THE BOUNDARY ELEMENT METHOD

A brief introduction of certain integrals is provided to facilitate an understanding of the nature of integrals resulting from a boundary element formulation. Consider the one-dimensional integral

$$I = \int_{L_1}^{L_2} \frac{f(r)}{|r|^\alpha} \, dr \qquad (6.1)$$

where $f(r)$ is a regular, integrable function between the limits L_1 and L_2, and α is a constant.

1. For $\alpha = 0$, the denominator in Eq. (6.1) is unity, the integrand, $f(r)$, is regular and the integral, I, is easily evaluated.
2. For $\alpha < 1$, the integrand, $f(r)/|r|^\alpha$, is singular at $r = 0$ but presents an integrable singularity. The integral, I, in this case is said to be *weakly singular*. For example, for $f(r) = 1$ and $\alpha = 0.5$,

$$I = \int_{L_1}^{L_2} \frac{1}{|r|^{0.5}} \, dr = \frac{1}{2} |r|^{0.5} \Bigg|_{L_1}^{L_2} \qquad (6.2)$$

The right-hand side expression in Eq. (6.2) is easily evaluated.

3. For $\alpha = 1$, the integrand, $f(r)/|r|$, is singular at $r = 0$ and the integral, I, may only be evaluated in the Cauchy principal value (CPV) sense (Churchill and Brown, 1978). The integral, I, is said to be *strongly singular*. From Fig. 6.2,

$$I = \int_{L_1}^{L_2} \frac{f(r)}{|r|} \, dr = \lim_{\varepsilon \to 0} \Bigg[\int_{L_1}^{-\varepsilon} \frac{f(r)}{|r|} \, dr$$
$$+ \int_{C} \frac{f(r)}{\varepsilon} \, ds + \int_{\varepsilon}^{L_2} \frac{f(r)}{|r|} \, dr \Bigg] \qquad (6.3)$$

where C is the semicircular contour around the singularity at $r = 0$. Noting that $ds = \varepsilon \, d\theta$ and $0 \le \theta \le \pi$, the singularity in the second integral is removed as ε cancels out in both the denominator and the numerator. Further, if the effects due to $-\varepsilon$ in the upper limit and $+\varepsilon$ in the lower limit in Eq. (6.3) cancel out, I is said to be integrable in the CPV sense.

4. For $\alpha > 1$, the integrand, $f(r)/|r|^{\alpha}$, is singular at $r = 0$ and cannot be evaluated in the CPV sense. The integral, I, in this case is said to be *hypersingular.*

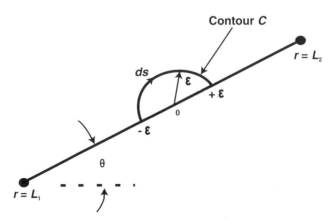

Figure 6.2 Singular integral domain in one dimension.

Consider next a two-dimensional integral of the form

$$I = \int_A \frac{f(x,\, y)}{|r|^\alpha} \, dA \tag{6.4}$$

where $f(x,\, y)$ is a regular function, $r = [(x - x_0)^2 + (y - y_0)^2]^{0.5}$ is the radial distance from an origin $(x_0,\, y_0)$, A is the domain under consideration containing the origin, and α is a constant.

1. For $0 < \alpha < 2$, a transformation to polar coordinates leads to

$$I = \int_r \int_\theta \frac{F(\theta)}{r^\alpha} \, r \, dr \, d\theta = \int_r \int_\theta \frac{F(\theta)}{r^{\alpha-1}} \, dr \, d\theta \tag{6.5}$$

where $F(\theta)$ is obtained from $f(x,\, y)$ after performing the transformation substitutions. The integration in Eq. (6.5) for the variable θ can be carried out in a straightforward manner. From the discussion above on one-dimensional integrals, Eq. (6.4) for variable r is integrable for $\alpha - 1 < 1$ or $\alpha < 2$.

2. For $\alpha = 2$, the integral in Eq. (6.4), using Fig. 6.3, can be written as

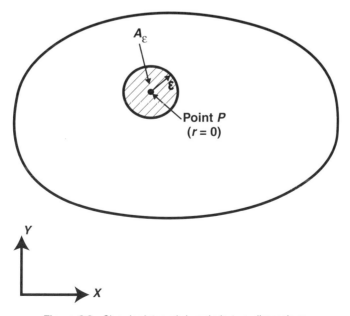

Figure 6.3   Singular integral domain in two dimensions.

$$I = \int_A \frac{f(x, y)}{r^2} \, dA = \lim_{A_\varepsilon \to 0} \left[\int_{A - A_\varepsilon} \frac{f(x, y)}{r^2} \, dA + \int_{A_\varepsilon} \frac{f(x, y)}{r^2} \, dA \right]$$

(6.6)

where A_ε is an infinitesimal area around the singularity at $r = 0$. The second integral on the right-hand side in Eq. (6.6) contains the singular term. If $f(x, y)$ satisfies certain properties, the integral, I, would still remain finite in the CPV sense.

3. For $\alpha > 2$, following the discussion on singular integrals in one dimension, the integral, I, in Eq. (6.4) cannot be evaluated in the CPV sense, and it is termed as *hypersingular*.

In three dimensions we consider the integral

$$I = \int_V \frac{f(x, y, z)}{r^\alpha} \, dV \qquad (6.7)$$

where $r = [(x - x_0)^2 + (y - y_0)^2 + (z - z_0)^2]^{0.5}$, α is a constant, and V is the volume domain under consideration, including the origin at which $r = 0$. Following arguments similar to those presented above for the one- and two-dimensional integrals and using the transformation $dV = r^2 \sin\phi \, dr \, d\phi \, d\theta$, where θ and ϕ are the spherical coordinate angles, respectively, we can deduce that

1. For $0 < \alpha < 3$, the singularity is integrable, the integral I in Eq. (6.7) is finite and is termed *weakly singular.*
2. For $\alpha = 3$, the integral, I, in Eq. (6.7) is finite provided $f(x, y, z)$ satisfies certain conditions and it can be evaluated in the CPV sense. The integral, I, is termed *strongly singular.*
3. For $\alpha > 3$, the integral, I, in Eq. (6.7) *is hypersingular* and cannot be evaluated in the CPV sense.

6.3 EQUATIONS OF ELASTICITY

The boundary element method is developed in this chapter with reference to elastic behavior of objects. It is advantageous, then, to present a brief outline of the equations of elasticity (Boresi and Chong, 2000).

The stress components, σ_{ij}, must satisfy the differential equations of equilibrium in a domain, Ω, under consideration:

$$\sigma_{ij,j} + F_i = 0 \qquad (6.8)$$

where F_i are the components of the body force acting on the object. The field equations above are to be solved given certain boundary conditions as

$$u_i = \bar{u}_i \qquad \text{on } \Gamma_u$$
$$t_i = \bar{t}_i \qquad \text{on } \Gamma_t \qquad (6.9)$$

where the tractions on a surface with normal n_j are given as $t_i = \sigma_{ij}n_j$ and $\bar{u}_i$ and $\bar{t}_i$ are the prescribed values of displacements and tractions on boundaries Γ_u and Γ_t, respectively. $\Gamma_u \cup \Gamma_t = \Gamma$ is the boundary of the subject domain Ω. The stress components, σ_{ij}, in Eq. (6.8) are symmetric (i.e., $\sigma_{ij} = \sigma_{ji}$). The kinematic equations, relating the strain components, ε_{ij}, to the displacements, u_i, are

$$\varepsilon_{ij} = \tfrac{1}{2}(u_{i,j} + u_{j,i}) \qquad (6.10)$$

where a comma (,) denotes differentiation, thus, $u_{i,j} = \partial u_i/\partial x_j$. The constitutive relations for a linear elastic material are given as

$$\sigma_{ij} = \lambda\varepsilon_{kk}\delta_{ij} + 2\mu\varepsilon_{ij} = E_{ijkl}\varepsilon_{kl} \qquad (6.11)$$

where λ and μ are the Lamé constants and E_{ijkl} is the constitutive tensor. Equations (6.8)–(6.11) summarize the equations of elastostatics for a linear, isotropic, homogenous elastic object.

6.4   FUNDAMENTAL OR KELVIN'S SOLUTION

The *fundamental solution,* also called the *Kelvin solution* (Love, 1944; Sokolnikoff, 1956), is now introduced. This solution is employed in the boundary element formulation to extract the solution of displacement components from the integral equations of the elastic body. The fundamental solution refers to the solution of the response of a linear, elastic solid of infinite extent due to the application of a point load.

Consider a domain of infinite extent shown in Fig. 6.4. A unit load is applied at the source point P, and the response at the field point Q is sought. Denoting all quantities related to the fundamental solution with an asterisk (*), the body force

$$F_i^* = \Delta(P, Q)\delta_{ij}e_j \qquad (6.12)$$

is substituted in Eq. (6.8) to obtain the fundamental solution. Here

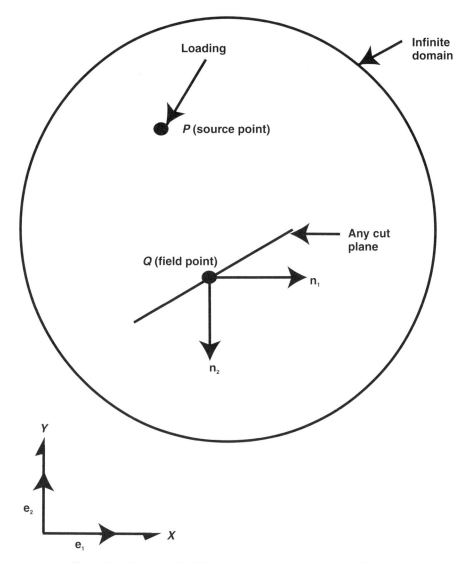

Figure 6.4 Domain of infinite extent under a concentrated point load.

$\Delta(P, Q)$ is the Dirac delta function, δ_{ij} is the Kronecker delta, and e_j is the unit vector in direction j. The fundamental (Kelvin) solution for displacement is then expressed symbolically as

$$u_i^* = U_{ij}^*(P, Q)e_j \qquad (6.13)$$

Substituting Eq. (6.13) into Eq. (6.10) followed by Eq. (6.11) and using the relation $t_i = \sigma_{ij}n_j$, the fundamental (Kelvin) solution for traction is obtained and is expressed here as

$$t_i^* = T_{ij}^*(P, Q)e_j \qquad (6.14)$$

For a two-dimensional body, these solutions are given as

$$U_{ij}^*(P, Q) = \frac{-1}{8\pi\mu(1 - \nu)} \left[(3 - 4\nu)\delta_{ij} \ln r - \frac{y_i y_j}{r^2} \right] \qquad (6.15)$$

$$T_{ij}^*(P, Q) = \frac{1}{4\pi(1 - \nu)r^2} \left\{ (1 - 2\nu)(n_j y_i - n_i y_j) \right.$$

$$\left. + \left[(1 - 2\nu)\delta_{ij} + \frac{2y_i y_j}{r^2} \right] y_k n_k \right\} \qquad (6.16)$$

where $r = y_i y_i$; $y_i = x_{pi} - x_{qi}$; x_{pi} is the x_i coordinate of the source point P and similarly for x_{qi}.

It is important here to understand the notation for the subscripts and for the arguments of the displacement and traction fundamental solutions. Thus, $U_{ij}^*(P, Q)$ denotes displacement at the location Q in the j direction due to a unit load applied at location P in the i direction. Similarly, $T_{ij}^*(P, Q)$ denotes the traction at location Q on an orientation defined by the normal n_j due to a unit load applied at location P in the i direction. It is easy, then, to see that $U_{ij}^*(P, Q)$ and $T_{ij}^*(P, Q)$ can each be written as 3×3 matrices for three-dimensional analyses and 2×2 matrices for two-dimensional analyses.

6.5 BOUNDARY ELEMENT FORMULATION

A starting point for the boundary element formulation is to multiply the equations of equilibrium in Eq. (6.8) by an arbitrary weighing function, $\overline{w}_i$ and integrate over the region considered:

$$\int_\Omega (\sigma_{ij,j} + F_i)\overline{w}_i \, d\Omega = 0 \qquad (6.17)$$

Noting that $(\sigma_{ij}\overline{w}_i)_{,j} = \sigma_{ij,j}\overline{w}_i + \sigma_{ij}\overline{w}_{i,j}$ from the product rule of differentiation (Stewart, 1999), the first term in the integral in Eq. (6.17) can be rewritten as

$$\int_\Omega [(\sigma_{ij}\overline{w}_i)_{,j} - \sigma_{ij}\overline{w}_{i,j}] \, d\Omega + \int_\Omega F_i\overline{w}_i \, d\Omega = 0 \qquad (6.18)$$

The first term in Eq. (6.18) is now rewritten using the divergence theorem of calculus (Schey, 1973). The *divergence theorem* for a vector field with components, R_j, is written as

$$\int_\Omega R_{j,k} \, d\Omega = \int_\Gamma R_j n_k \, d\Gamma \qquad (6.19)$$

where Γ is the bounding surface for the volume Ω and n_k is the outward normal to the surface $d\Gamma$. Substituting $R_j = \sigma_{ij}\overline{w}_i$ and replacing the index k by j in Eq. (6.19), the first term under the integral in Eq. (6.18) can be written as

$$\int_\Omega (\sigma_{ij}\overline{w}_i)_{,j} \, d\Omega = \int_\Gamma \sigma_{ij}\overline{w}_i n_j \, d\Gamma \qquad (6.20)$$

Since $\sigma_{ij}n_j = t_i$, this term is further simplified as

$$\int_\Omega (\sigma_{ij}\overline{w}_i)_{,j} \, d\Omega = \int_\Gamma t_i\overline{w}_i \, d\Gamma \qquad (6.21)$$

We now turn our attention to the second integral in Eq. (6.18). Since the stress tensor σ_{ij} is symmetrical ($\sigma_{ij} = \sigma_{ji}$), the integrand for this term may be written as

$$\sigma_{ij}\overline{w}_{i,j} = \tfrac{1}{2}\sigma_{ij}(\overline{w}_{i,j} + \overline{w}_{j,i}) \qquad (6.22)$$

Using the constitutive relation $\sigma_{ij} = E_{ijkl}\varepsilon_{kl}$ from Eq. (6.11) and denoting $\tfrac{1}{2}(\overline{w}_{i,j} + \overline{w}_{j,i})$ as $\overline{\varepsilon}_{ij}$, by considering the weighing function $\overline{w}_i$ as a displacement field and employing the strain–displacement relationships of Eq. (6.10), we obtain

$$\sigma_{ij}\overline{w}_{i,j} = E_{ijkl}\varepsilon_{kl}\overline{\varepsilon}_{ij} \tag{6.23}$$

The strain field ε_{kl} is next expressed in terms of its displacement components as $\varepsilon_{kl} = \frac{1}{2}(u_{k,l} + u_{l,k})$. Also, using the symmetry of the constitutive tensor $E_{ijkl} = E_{klij}$ (Chung, 1988), it is combined with the strain field ε_{ij} to give $\overline{\sigma}_{kl} = E_{klij}\,\overline{\varepsilon}_{ij}$. The integrand under consideration is then obtained as

$$\sigma_{ij}\overline{w}_{i,j} = \frac{1}{2}(u_{k,l} + u_{l,k})\overline{\sigma}_{kl} \tag{6.24}$$

Finally, noting the symmetry of $\overline{\sigma}_{kl}$, (6.24) is written as

$$\sigma_{ij}\overline{w}_{i,j} = u_{k,l}\overline{\sigma}_{kl} \tag{6.25}$$

To allow use of the divergence theorem on this term, it is first rewritten using the product rule of differentiation. The second integral in Eq. (6.18) can then be written as

$$\int_{\Omega} \sigma_{ij}\overline{w}_{i,j}\, d\Omega = \int_{\Omega} [(\overline{\sigma}_{kl}u_k)_{,l} - \overline{\sigma}_{kl,l}u_k]\, d\Omega \tag{6.26}$$

Applying the divergence theorem to the first term on the right-hand side of Eq. (6.26), we obtain

$$\int_{\Omega} \sigma_{ij}\overline{w}_{i,j}\, d\Omega = \int_{\Gamma} \overline{\sigma}_{kl}u_k n_l\, d\Gamma - \int_{\Omega} \overline{\sigma}_{kl,l}u_k\, d\Omega \tag{6.27}$$

Expressing $\overline{\sigma}_{kl}n_l = \overline{t}_k$ and noting that the stress field $\overline{\sigma}_{kl}$ satisfies the equilibrium relations in Eq. (6.8) for a body force field $\overline{F}_k$ (i.e., $\overline{\sigma}_{kl,l} + \overline{F}_k = 0$), Eq. (6.27) is rewritten as

$$\int_{\Omega} \sigma_{ij}\overline{w}_{i,j}\, d\Omega = \int_{\Gamma} \overline{t}_k u_k\, d\Gamma + \int_{\Omega} \overline{F}_k u_k\, d\Omega \tag{6.28}$$

Substituting Eqs. (6.21) and (6.28) into (6.18) and upon rearranging terms, we get

$$\int_{\Gamma} t_i\overline{w}_i\, d\Gamma + \int_{\Omega} F_i\overline{w}_i\, d\Omega = \int_{\Gamma} \overline{t}_i u_i\, d\Gamma + \int_{\Omega} \overline{F}_i u_i\, d\Omega \tag{6.29}$$

Equation (6.29) is the familiar *Betti's reciprocal theorem* of structural analysis (Norris et al., 1976). Several boundary element formulations employ Eq. (6.29) as their starting point.

The weighting function, $\overline{w}_i$, in our discussions above is an arbitrary function that has taken on the characteristics of a displacement field. The corresponding stress, strain, traction, and body force fields are denoted as $\overline{\sigma}_{ij}$, $\overline{\varepsilon}_{ij}$, $\overline{t}_i$, and $\overline{F}_i$, respectively. We now select the weighting function to be the fundamental displacement solution, the displacement field for an infinite elastic body under a unit point load, as outlined in Section 6.4. Thus

$$\overline{F}_i = F_i^* = \Delta(P, Q)\delta_{ij}e_j \tag{6.30a}$$

$$\overline{w}_i = u_i^* = U_{ij}^*(P, Q)e_j \tag{6.30b}$$

$$\overline{t}_i = t_i^* = T_{ij}^*(P, Q)e_j \tag{6.30c}$$

Substituting these relations into Eq. (6.29), and canceling e_j in each term of the resulting equation, we get

$$\int_\Gamma t_i(Q)U_{ij}^*(P, Q)\, d\Gamma + \int_\Omega F_i(Q)U_{ij}^*(P, Q)\, d\Omega$$

$$= \int_\Gamma u_i(Q)T_{ij}^*(P, Q)\, d\Gamma + \int_\Omega u_i(Q)\, \Delta(P, Q)\delta_{ij}\, d\Omega \tag{6.31}$$

Equation (6.31) is the well-known *Somigliana's identity* (Somigliana, 1886) and has also been used as the starting point for several boundary element formulations.

To continue the boundary element formulation, we consider the second term on the right-hand side of Eq. (6.31) in further detail. The quantity $\Delta(P, Q)$ in the integrand vanishes everywhere except at P. The integral, using Fig. 6.5, may be written as

$$\int_\Omega u_i(Q)\, \Delta(P, Q)\delta_{ij}\, d\Omega = \int_{\Omega-\Omega_\varepsilon} u_i(Q)\, \Delta(P, Q)\delta_{ij}\, d\Omega$$

$$+ \int_{\Omega_\varepsilon} u_i(Q)\, \Delta(P, Q)\delta_{ij}\, d\Omega \tag{6.32}$$

where Ω_ε is a sphere of radius ε surrounding the singularity at $r = 0$. The first term on the right-hand side of Eq. (6.32) vanishes as $\Delta(P, Q)$ is zero everywhere in $\Omega - \Omega_\varepsilon$. The second term is written as

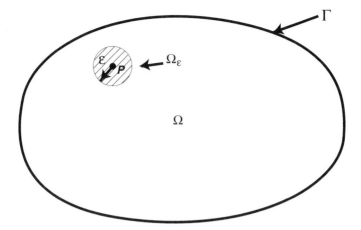

Figure 6.5 Domain of an elastic body and the source point.

$$\int_{\Omega_\varepsilon} u_i(Q)\ \Delta(P,\ Q)\delta_{ij}\ d\Omega\ =\ \varepsilon\delta_{ij}u_i(P) \tag{6.33}$$

where $\varepsilon = 0, \frac{1}{2}$, and 1, depending on whether P lies outside Γ, is within the volume Ω, or lies on a smooth boundary Γ, respectively (see Section 6.15). The term on the right-hand side of Eq. (6.33) is called the *free term*. Our system, at this point, is represented by Eq. (6.31). Considering the relation written in Eq. (6.33), the only term containing a volume integration in Eq. (6.31) is the one due to the body force $F_i(Q)$. To obtain a representation that involves only the surface integrals, we consider the cases for which the body forces, $F_i(Q)$, are absent in Eq. (6.31). Using Eq. (6.33) and $F_i(Q) = 0$, with a rearrangement of terms, Eq. (6.31) reduces to

$$\varepsilon\delta_{ij}u_i(P) = \int_\Gamma [t_i(Q)U_{ij}^*(P,\ Q) - u_i(Q)T_{ij}^*(P,\ Q)]\ d\Gamma(Q) \tag{6.34}$$

Equation (6.34) is the basic boundary integral equation that forms the basis for the boundary element method. In the form given in Eq. (6.34), no volume integrals are involved and only surface integrals need be evaluated. The order of the system in going from volume to surface integrals (and from surface to line integrals in two dimensions) has been reduced by one. This feature has constituted a major attraction of the boundary element method in the area of numerical analysis methods.

It is recognized here that the body forces have been neglected in obtaining the boundary-only representation of Eq. (6.34). The inclusion of such forces at this point would lead to a volume integral and destroy the boundary-only nature of Eq. (6.34). Various treatments of body forces that preserve the boundary-only nature of the boundary integral formulation are presented in Sections 6.12 and 6.13.

6.6   DISPLACEMENT AND TRACTION INTERPOLATION

The boundary element method is frequently referred to as a *mixed formulation,* as it involves both the displacements as well as the tractions as unknown variables. Interpolation functions, similar to those described in Chapter 4, are required to describe the variation of displacements and tractions, respectively, over an element in terms of their nodal values at the nodes of the element. A one-dimensional boundary element suitable for use in two-dimensional analysis is shown in Fig. 6.6. A quadratic three-node element is considered here. The formulations for constant and linear elements follow along similar lines. The interpolations for the unknown variables, namely displacements, u, and tractions, t, are written using the shape functions, N_i, as

$$u(\xi) = \sum_{i=1}^{3} N_i(\xi)u_i \qquad (6.35a)$$

$$t(\xi) = \sum_{i=1}^{3} N_i(\xi)t_i \qquad (6.35b)$$

where u_i and t_i are the displacement and traction, respectively, at the node i of the boundary element.

The shape functions, $N_i(\xi)$, are given as

$$N_1(\xi) = 2(\xi - 0.5)(\xi - 1)$$

$$N_2(\xi) = -4\xi(\xi - 1) \qquad (6.36)$$

$$N_3(\xi) = 2\xi(\xi - 0.5)$$

$\xi = 0$ $\xi = 0.5$ $\xi = 1.0$ ξ

Figure 6.6   One-dimensional quadratic boundary element: •, end nodes; ×, midside node.

Nodes 1, 2, and 3 of the quadratic element lie at the nondimensional locations ξ = 0, 0.5, and 1.0, respectively. It may be verified that the shape functions satisfy their usual properties. The shape function N_1 = 1 at node 1 and vanishes at nodes 2 and 3, respectively. Shape functions N_2 and N_3 behave similarly. Additionally, by Eq. (6.36), $\sum_{i=1}^{3} N_i(\xi)$ = 1. The shape functions of Eq. (6.36) are illustrated graphically in Fig. 6.7. Equations (6.35a) and (6.35b) are written in the matrix notation as

$$\{u\} = [N]\{u_i\} \tag{6.37a}$$

$$\{t\} = [N]\{t_i\} \tag{6.37b}$$

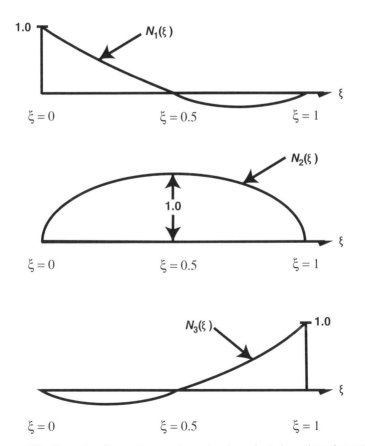

Figure 6.7 Shape functions of a one-dimensional quadratic boundary element.

where

$$\{u\} = \begin{Bmatrix} u_x \\ u_y \end{Bmatrix} \qquad \{t\} = \begin{Bmatrix} t_x \\ t_y \end{Bmatrix} \qquad (6.37c)$$

$$[N] = \begin{bmatrix} N_1 & 0 & N_2 & 0 & N_3 & 0 \\ 0 & N_1 & 0 & N_2 & 0 & N_3 \end{bmatrix} \qquad (6.37d)$$

$$\{u_i\} = \begin{Bmatrix} u_x^1 \\ u_y^1 \\ u_x^2 \\ u_y^2 \\ u_x^3 \\ u_y^3 \end{Bmatrix} \qquad \{t_i\} = \begin{Bmatrix} t_x^1 \\ t_y^1 \\ t_x^2 \\ t_y^2 \\ t_x^3 \\ t_y^3 \end{Bmatrix} \qquad (6.37e)$$

where, for example, u_x^1 in displacement vector $\{u_i\}$ is the displacement at element node 1 along the x direction and similarly for terms in the traction vector $\{t_i\}$.

6.7 ELEMENT CONTRIBUTIONS

We seek to evaluate the surface integral in Eq. (6.34) numerically and develop a matrix form of that equation. The matrix relation so developed may be solved to obtain the unknown displacements and tractions at the nodes of the boundary mesh. To evaluate the surface integral, it is first expressed as the sum of integrals over the boundary of each element. Thus,

$$\int_\Gamma (\cdot)\, d\Gamma = \sum_{n=1}^{\text{NEL}} \int_{\Gamma_n} (\cdot)\, d\Gamma \qquad (6.38)$$

where NEL represents the number of elements used to discretize the boundary and Γ_n is the boundary of the boundary element n. The summation above is performed by assembling the matrix contributions from each element. The assembly process is explained in Section 6.8. The procedure for evaluation of the contribution from each element is now outlined. The first term on the right-hand side of Eq. (6.34) for a single element appears as

$$\alpha_j = \int_{\Gamma_n} t_i(Q) U_{ij}^*(P, Q) \, d\Gamma(Q) \tag{6.39}$$

As explained previously, for two-dimensional analysis the fundamental solution $U_{ij}^*(P, Q)$ may be written as a 2×2 matrix. Equation (6.39) can then be expressed in matrix form as

$$\{\alpha\} = \int_{\Gamma_n} [U^*]\{t\} \, d\Gamma(Q) \tag{6.40a}$$

where

$$\{\alpha\} = \begin{Bmatrix} \alpha_1 \\ \alpha_2 \end{Bmatrix} \tag{6.40b}$$

and

$$[U^*] = \begin{bmatrix} U_{11}^* & U_{21}^* \\ U_{12}^* & U_{22}^* \end{bmatrix} \tag{6.40c}$$

Substituting from Eq. (6.37b), Eq. (6.40a) can be written in terms of element nodal tractions as

$$\{\alpha\} = \int_{\Gamma_n} [U^*][N]\{t_i\} \, d\Gamma(Q) \tag{6.41}$$

Since $\{t_i\}$ are nodal tractions, they can be taken outside the integral and Eq. (6.41) can be written in compact matrix form as

$$\{\alpha\} = [G_n]\{t_i\} \tag{6.42}$$

with

$$[G_n] = \int_{\Gamma_n} [U^*][N] \, d\Gamma(Q) \tag{6.43}$$

where $[G_n]$ is a (2×6) matrix and is the contribution of boundary element n to the first term on the right-hand side of Eq. (6.34). The contribution of this boundary element to the second term on the right-hand side of Eq. (6.34) may similarly be written as

$$\{\beta\} = [F_n]\{u_i\} \tag{6.44}$$

with

$$[F_n] = \int_{\Gamma_n} [T^*][N] \, d\Gamma(Q) \tag{6.45}$$

where $[F_n]$ is a (2×6) matrix contribution due to element n. The expression in Eq. (6.34) can now be rewritten after these substitutions and that from Eq. (6.38) as

$$\varepsilon \delta_{ij} u_i(P) = \sum_{n=1}^{\text{NEL}} \int_{\Gamma_n} t_i(Q) U_{ij}^*(P, Q) \, d\Gamma(Q)$$

$$- \sum_{n=1}^{\text{NEL}} \int_{\Gamma_n} u_i(Q) T_{ij}^*(P, Q) \, d\Gamma(Q) \tag{6.46}$$

and then expressed in the discretized form as

$$\varepsilon \delta_{ij} u_i(P) = \sum_{n=1}^{\text{NEL}} \int_{\Gamma_n} [U^*][N] \, d\Gamma(Q)\{t_i\}$$

$$- \sum_{n=1}^{\text{NEL}} \left(\int_{\Gamma_n} [T^*][N] \, d\Gamma(Q) \right)\{u_i\} \tag{6.47}$$

Equation (6.47) now deserves a closer inspection. It is an expression for the displacement at a point P which may lie anywhere on the body either within its domain Ω or on its bounding surface Γ. The displacement $u_i(P)$ is expressed in terms of the displacements $\{u_i\}$ and the tractions $\{t_i\}$ at nodes that lie on the boundary Γ. In the case of two-dimensional analysis, a discretization of the boundary using N nodes will lead to $2N$ displacement and $2N$ traction quantities from which the displacement $u_i(P)$ at point P can be found. It is realized further that, in general, a total of $2N$ of these quantities, either tractions or displacements, are known from the boundary conditions prescribed. Thus, there are a total of $2N$ unknowns on the right-hand side of Eq. (6.47) for which we require $2N$ independent equations. Recall again that the source point P in Eq. (6.47) may lie anywhere on the object, including its boundary. Taking advantage of this fact, the source point P is located successively at each nodal point on the boundary Γ. For each location of P on a boundary node, two independent equations, one corresponding to each coordinate direction, are obtained. Thus, the required $2N$ independent equations are obtained by considering the application of Eq. (6.47) successively at each nodal point.

Consider now the load point P placed at node j on the boundary. The displacements $u_1(P)$ and $u_2(P)$ obtained from Eq. (6.47) correspond to the displacements of node j in the x- and y-directions, respectively. These unknown displacements at node j also appear on the right-hand side of Eq. (6.47) when considering integration over element(s) for which node j is either an end node or a midside node. In the assembly process discussed in Section 6.8, the coefficients of the unknown displacements coming from the left- and right-hand sides of Eq. (6.47) are all lumped together. It is noted further that an interesting case arises when the load point P is at the node point j and the contributions to the surface integral on the right-hand side of Eq. (6.47) from an element that contains node j are considered. In this case, the quantity r, the distance between the source point and the field point, approaches zero. Since r appears in the fundamental solutions, which are included in the integrands in Eq. (6.47) the integrals are of a singular nature. For two-dimensional analysis, the singularities are due to an r term present in the displacement solution U_{ij}^* and a $1/r$ term present in the traction solution T_{ij}^*. Referring to the discussion on singular one-dimensional integrals in Section 6.2, the $\ln r$ term presents a weak singularity and the integrals can still be evaluated, albeit with care. The $1/r$ term, on the other hand, presents a strong singularity and cannot easily be evaluated. Fortunately, these strongly singular terms can be evaluated directly without having to calculate the singular integrals as shown in Sections 6.9 and 6.14.

6.8 ASSEMBLY OF BOUNDARY ELEMENT MATRICES

The contributions $[G_n]$ and $[F_n]$ of the element n with boundary Γ_n were obtained in Eqs. (6.43) and (6.45), respectively. Assuming that the boundary Γ of the object is smooth and without any sharp corners, the tractions at a node shared by two neighboring elements are continuous, otherwise they are not. Regardless of whether the boundary is smooth or not, the displacements at such shared nodes are always continuous. The continuity of displacements at shared nodes of two neighboring elements is exploited to perform the assembly of boundary element matrices. The assembly process for the $[G_n]$ and $[F_n]$ contributions arising from three neighboring elements $(n - 1)$, n, and $(n + 1)$ is shown in Fig. 6.8 corresponding to the case when the load point, P, is located at node j. The double-hatched areas denote the assembly of two matrices where the contributions from these matrices are added to each other. These areas contain contributions corresponding to the node shared by two adjacent elements. A boundary discretization containing N nodes in a two-dimensional analysis will therefore lead to an assembled $[F]$ matrix of size $2N \times 2N$. The assembly of the matrices $[G_n]$ cannot be performed similarly, as the tractions at nodes shared by elements may not be continuous. One strategy for the assembly of these matrices is shown in Fig. 6.9. The contributions arising from the neighboring elements $(n - 1)$, n, and $(n + 1)$ are simply placed next to each other. For a boundary element discretization consisting of N nodes and NEL elements in a two-dimensional analysis, the assembled $[G]$ matrix is of size $2N \times 6NEL$. The resulting $[G]$ matrix is rectangular and therefore cannot be inverted.

Upon completing the assembly process and absorbing the $\varepsilon \delta_{ij}$ coefficient term with the corresponding coefficients for $u_i(P)$ on the right-hand side, Eq. (6.47) is written in matrix form as

$$[F]\{u\} = [G]\{t\} \qquad (6.48)$$

A solution of Eq. (6.48) must be performed to obtain the unknown displacements and tractions at the boundary.

6.9 RIGID-BODY MOTION

Before performing the solution of the unknowns in Eq. (6.48), attention is focused again on the elements of matrix $[F]$. For a two-dimensional analysis, the 2×2 diagonal matrix elements of this matrix need to be obtained by computing strongly singular integrations. As pointed out in Section 6.7, these terms arise when the load point P is at the node

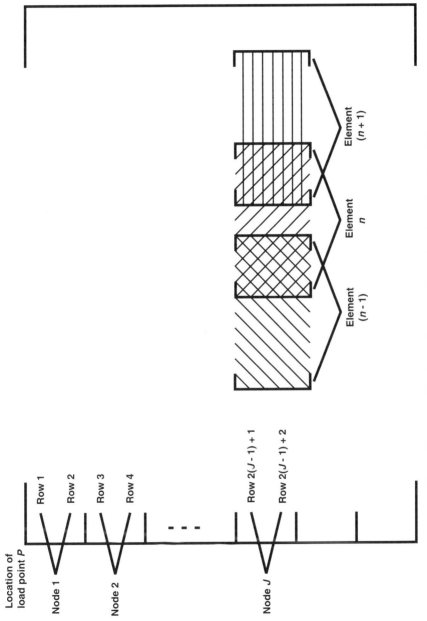

Figure 6.8 Assembly of two-dimensional boundary element matrices $[F_n]$.

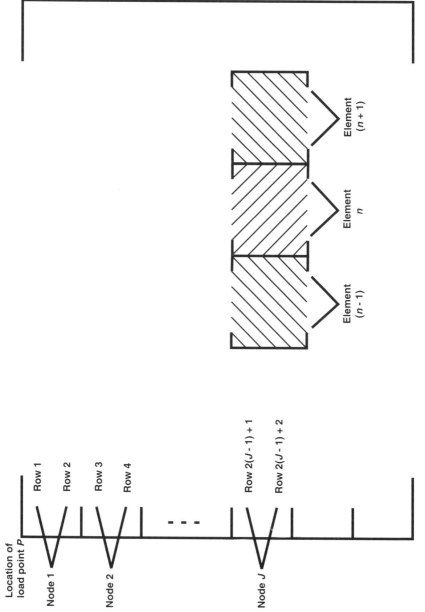

Figure 6.9 Assembly of two-dimensional boundary element matrices $[G_n]$.

point j and the contributions to the surface integral on the right-hand side of Eq. (6.47) from an element that contains node j are considered. In this case, the quantity r, the distance between the source point and field point, approaches zero. Since r appears in the fundamental solutions, which are included in the integrands in Eq. (6.47) the integrals are of a singular nature. For two-dimensional analysis, the singularities are due to an ln r term present in the displacement solution U_{ij}^* and an $1/r$ term present in the traction solution T_{ij}^*. Referring to the discussion on singular one-dimensional integrals in Section 6.2, the ln r term presents a weak singularity and the integrals can still be evaluated, albeit with care. The $1/r$ term, on the other hand, presents a strong singularity and cannot be easily evaluated. A numerical procedure for such integrals is not desirable, as it may require excessive computational effort and still lead to erroneous results. The off-diagonal terms are, however, all easily computed using appropriate numerical integration schemes. Fortunately, the integrals for strongly singular 2×2 matrix contributions do not need to be computed and can be obtained using the rigid-body motion property of solid objects. This property states that a rigid-body motion of the object does not lead to the development of tractions on the body. Consider the two rigid-body motions given as

$$
\{u\} = \begin{Bmatrix} 1 \\ 0 \\ 1 \\ 0 \\ \vdots \\ \vdots \\ 1 \\ 0 \end{Bmatrix} \qquad \{u\} = \begin{Bmatrix} 0 \\ 1 \\ 0 \\ 1 \\ \vdots \\ \vdots \\ 0 \\ 1 \end{Bmatrix} \tag{6.49}
$$

In Eq. (6.49), the first $\{u\}$ vector denotes a rigid-body motion of magnitude unity in the x direction, while the second $\{u\}$ vector denotes a rigid-body motion of magnitude unity in the y direction. Under either of these rigid-body motions, the traction vector $\{t\} = \{0\}$, making the right-hand side of Eq. (6.48) equal to zero. Substituting this result in Eq. (6.48) first for the case of unit rigid-body motion in the x direction, we obtain

$$
\begin{bmatrix}
F_{11} & F_{12} & \cdots & \cdots & \cdots & \cdots & \cdots & F_{1,2N} \\
\cdots & \cdots & \cdots & \cdots & \cdots & \cdots & \cdots & \cdots \\
\cdots & \cdots & \cdots & \cdots & \cdots & \cdots & \cdots & \cdots \\
F_{k,1} & F_{k,2} & \cdots & F_{k,k} & F_{k,k+1} & \cdots & \cdots & F_{k,2n} \\
\cdots & \cdots & \cdots & \cdots & \cdots & \cdots & \cdots & \cdots \\
\cdots & \cdots & \cdots & \cdots & \cdots & \cdots & \cdots & \cdots \\
\cdots & \cdots & \cdots & \cdots & \cdots & \cdots & \cdots & \cdots \\
F_{2N,1} & F_{2N,2} & \cdots & \cdots & \cdots & \cdots & \cdots & F_{2N,2N}
\end{bmatrix}
\begin{Bmatrix}
1 \\ 0 \\ \vdots \\ 1 \\ 0 \\ \vdots \\ 1 \\ 0
\end{Bmatrix}
=
\begin{Bmatrix}
0 \\ 0 \\ 0 \\ 0 \\ 0 \\ 0 \\ 0 \\ 0
\end{Bmatrix}
\quad (6.50)
$$

where for $k = 2(j - 1) + 1$, F_{kk} and $F_{k,k+1}$ represent two of the four singular terms in the 2×2 diagonal submatrix arising from considering the load point P to lie on node j. The other two singular terms appear in the equation for the row $k + 1$. Expanding row k of Eq. (6.50) and rearranging yields

$$
F_{k,k} = -(F_{k,1} + F_{k,3} + \cdots + F_{k,2N-1}) \quad (6.51)
$$

The term $F_{k,k+1}$ can be similarly obtained by considering the unit rigid-body motion in the y direction. The singular diagonal terms can thus be computed as the negative sum of the off-diagonal terms and do not need to be computed from the corresponding singular integrals. This property also leads to the diagonal dominance of the assembled $[F]$ matrix.

6.10 SOLUTION OF BOUNDARY ELEMENT EQUATIONS

The system of assembled boundary element equations in Eq. (6.48) is solved to obtain the unknown displacements and tractions at the boundary. The matrices $\{u\}$ and $\{t\}$ in Eq. (6.48) each contain both the known (through prescribed boundary conditions) and the unknown respective components. The system shown in Eq. (6.48) is partitioned as

$$
\begin{bmatrix}
[F_{uu}] & [F_{uk}] \\
\hline
[F_{ku}] & [F_{kk}]
\end{bmatrix}
\begin{Bmatrix}
\{u_u\} \\
\{u_k\}
\end{Bmatrix}
=
\begin{bmatrix}
[G_{uu}] & [G_{uk}] \\
\hline
[G_{ku}] & [G_{kk}]
\end{bmatrix}
\begin{Bmatrix}
\{t_u\} \\
\{t_k\}
\end{Bmatrix}
\quad (6.52)
$$

The subscripts u and k in Eq. (6.52) denote unknown and known components, respectively. Thus, $\{u_k\}$ and $\{t_k\}$ are the known nodal displacements and tractions obtained from the boundary conditions pre-scribed. Similarly, $\{u_u\}$ and $\{t_u\}$ are the unknown nodal displacements and tractions. Expanding the matrix equations in (6.52) and rearranging to have all unknowns on the left-hand side, we obtain

$$\left[\begin{array}{c|c} [F_{uu}] & [-G_{uu}] \\ \hline [F_{ku}] & [-G_{ku}] \end{array}\right]\left\{\begin{array}{c} \{u_u\} \\ \{t_u\} \end{array}\right\} = \left[\begin{array}{c|c} [-F_{uk}] & [G_{uk}] \\ \hline [-F_{kk}] & [G_{kk}] \end{array}\right]\left\{\begin{array}{c} \{u_k\} \\ \{t_k\} \end{array}\right\} \qquad (6.53)$$

All terms on the right-hand side of Eq. (6.53) are known. After per-forming the matrix multiplication, the right-hand side can be replaced by a column matrix $\{b\}$. Equation (6.53) can then be written in the standard form for linear matrix equations as

$$[A]\{x\} = \{b\} \qquad (6.54)$$

where [A] is the square coefficient matrix on the left-hand side in Eq. (6.53) and contains contributions from both [F] and [G] matrices. The matrix $\{x\}$ contains all the unknown nodal displacements and tractions on the boundary for the system, and $\{b\}$ is a known matrix. The matrix [A] is fully populated and unsymmetric, as are the matrices [F] and [G] in Eq. (6.48). Equation (6.54) may be solved using a direct or an iterative solver for fully populated, unsymmetric matrices.

It is clear from Eq. (6.48) that the terms of the [F] and [G] matrices are not of comparable magnitude. The introduction of terms from both [F] and [G] matrices in matrix [A] may lead to ill-conditioning of matrix [A]. The solution of Eq. (6.54) may then lead to numerical difficulties as well as erroneous results. The contributions from the [G] matrix to the matrix [A] are commonly scaled as

$$\left[\begin{array}{c|c} [F_{uu}] & \alpha[-G_{uu}] \\ \hline [F_{ku}] & \alpha[-G_{ku}] \end{array}\right]\left\{\begin{array}{c} u_u \\ t_u/\alpha \end{array}\right\} = \{b\} \qquad (6.55)$$

where α is a scaling parameter. For elasticity problems, α is taken to be the shear modulus.

6.11   DISPLACEMENT AT POINTS IN THE INTERIOR

The solution of Eq. (6.54), or more precisely Eq. (6.55), yields the unknown displacements $\{u_u\}$ and the unknown tractions $\{t_u\}$. Together with the known $\{u_k\}$ and $\{t_k\}$ matrices, the displacements $\{u\}$ and $\{t\}$

on the entire boundary Γ are known. It is, however, often desirable to know the displacements at one or several points within the interior of the body. Recall that Eq. (6.47), with an appropriate value of ε, is valid both inside the body and on its boundary. For a point P inside the boundary, $\varepsilon = \frac{1}{2}$. Equation (6.47) can then be used directly to obtain the displacement at any interior location P. Since the load point P is inside the boundary and the field point Q is on the boundary, the quantity r in the fundamental solutions U_{ij}^* and T_{ij}^* in Eqs. (6.15) and (6.16) is never zero. Thus, no singular integrals are involved in calculating displacements at the interior points. The matrices $[U^*]$ and $[T^*]$ in Eq. (6.47) must be computed anew for each interior location P for which displacements are required. Thus, a significant computational effort is expended if the displacements at a large number of points within the interior are required. This is unlike the finite element method where the solution of the finite element matrix equations at once yields the displacements of all interior as well as boundary points of the object.

6.12 BODY FORCES

The body forces acting on the elastic object under consideration have been neglected so far to enable us to obtain a boundary-only system of integral equations. The contribution due to body forces (gravity and thermally induced forces, etc.) results in an additional volume integral term given as [from Eq. (6.31)]

$$\int_{\Omega} F_i(Q) U_{ij}^*(P, Q) \, d\Omega \tag{6.56}$$

To evaluate this volume integral the domain of the body must be discretized into integration cells. Equation (6.56) does not contain the unknown displacement and traction components of vector $\{x\}$ of Eq. (6.54). The contribution from this term augments the right-hand side matrix $\{b\}$ in Eq. (6.54). Thus, while the domain of the body is discretized into integrations cells (much like the finite elements), unlike the finite element method no additional unknowns corresponding to the interior of the body are introduced. Regardless, the appearance of the volume integral and its subsequent evaluation by discretizing the domain reduces the attractiveness of the boundary element method. Some of the limitations of the finite element method associated with meshing arbitrary regions are then believed by most analysts to apply also to the boundary element method. It is noted, however, that ill-shaped integration cells do not seriously affect the quality of results obtained using the boundary elements. On the other hand, finite elements of undesirable shape may lead to serious errors in analysis. Numerous efforts have been made in the literature to eliminate the domain integral due to body forces and replace it by equivalent surface integrals. Three of the prominent techniques in this regard are (1) the particular integral approach, (2) the dual reciprocity approach, and (3) the multiple reciprocity approach. These approaches are quite similar to each other. The particular integral approach appears to be quite straightforward and is explained next in further detail.

6.13   PARTICULAR INTEGRAL APPROACH

As stated in Section 6.12, the inclusion of the effect of body forces through a boundary-only formulation presents significant advantages for the boundary element method. The particular integral approach (Pape and Banerjee, 1987) provides a straightforward, yet highly general approach to achieve this objective. This approach merely involves nodal evaluations and does not even require the computation of boundary integrals. The equilibrium statement in Eq. (6.8) can be written in terms of displacements through the substitution of stress and strain relations given in Eqs. (6.10) and (6.11), respectively. The resulting equation, called *Navier's equation of elasticity,* is given as (Boresi and Chong, 2000)

$$(\lambda + \mu)u_{j,ij} + \mu u_{i,jj} + F_i = 0 \qquad (6.57a)$$

where λ and μ are the Lamé elastic coefficients. In the operator form Eq. (6.57a) may be written as

$$\Lambda(u) + F_i = 0 \qquad (6.57b)$$

where Λ is a linear differential operator. From standard calculus (Boyce and DiPrima, 1977), the solution u, to a nonhomogeneous differential equation may be expressed as the sum of a complementary solution, u^c, and a particular integral solution, u^p, satisfying

$$\Lambda(u^c) = 0 \qquad (6.58)$$

$$\Lambda(u^p) = -F_i \qquad (6.59)$$

with

$$u_i = u_i^c + u_i^p \tag{6.60}$$

The superscripts c and p denote complementary and particular solutions, respectively. Equation (6.58) is, indeed, the equilibrium equation with zero body force for which we earlier developed the boundary integral solution. The complementary solution u^c therefore satisfies the matrix relation in Eq. (6.48) and is written as

$$[F]\{u^c\} = [G]\{t^c\} \tag{6.61}$$

Equation (6.59) may now be solved directly to obtain the particular solution u^p. The particular solution may be obtained easily if the body forces are given in terms of polynomials. For nonuniform distributions of body forces, these may be approximated by polynomials using least squares (Saigal et al., 1990b), Fourier series (Zeng and Saigal, 1991), or other simple techniques. For several cases, the particular solution may even be obtained by inspection. It is noted here that the particular integral for u^p satisfying Eq. (6.59) is not unique. Any polynomial or a linear combination of polynomials satisfying Eq. (6.59) is a valid particular integral solution. Assuming that the particular integral solution is available, the complementary solution, u^c, is expressed by rearranging Eq. (6.60) and writing it in the matrix form as

$$\{u^c\} = \{u\} - \{u^p\} \tag{6.62}$$

Substituting Eq. (6.62) into (6.61) and rearranging terms leads to

$$[F]\{u\} = [G]\{t\} + \{f\} \tag{6.63}$$

where

$$\{f\} = [F]\{u^p\} - [G]\{t^p\} \quad \text{and} \quad \{t^p\} = \{t\} - \{t^c\} \tag{6.64}$$

Thus, the body forces using the particular integral approach lead merely to the addition of a force term, $\{f\}$, on the right-hand side of boundary element equations. The matrices $[F]$ and $[G]$ required for computing $\{f\}$ are already available. The particular solutions $\{u^p\}$ and $\{t^p\}$ required to compute $\{f\}$ are direct nodal evaluations and do not require the computations of any volume or boundary integrals.

To illustrate the computation of particular integral solution, consider the common case of gravity loading due to self-weight. This loading is expressed mathematically for two-dimensional analysis as

$$F_1 = 0 \qquad F_2 = -\gamma \tag{6.65}$$

where γ is the weight density. Substituting Eq. (6.65) into Navier's equation of elasticity (6.57), the solution, u^p, using the Galerkin vector (Boresi and Chong, 2000, Chap. 8) can be written as

$$u_1^p = \frac{-\lambda}{4\mu(\lambda + \mu)} \gamma x_1 x_2 \tag{6.66}$$

$$u_2^p = \frac{1}{8\mu(\lambda + \mu)} \gamma[(\lambda + 2\mu)^2 x_2^2 + \lambda x_1^2] \tag{6.67}$$

Substituting the displacement relations in Eqs. (6.66) and (6.67) into the strain–displacement relations of Eq. (6.10) and thence into the stress–strain relations of Eq. (6.11), the stresses due to self-weight are obtained as

$$\sigma_{11}^p = \sigma_{12}^p = 0 \qquad \sigma_{22}^p = \gamma x_2 \tag{6.68}$$

Using the relation $t_i = \sigma_{ij} n_j$, the corresponding tractions are obtained as

$$t_1^p = 0 \qquad t_2^p = \gamma x_2 n_2 \tag{6.69}$$

where x_1 and x_2 are the coordinates of the point where these traction quantities are being evaluated. Thus, for self-weight body forces the particular solutions $\{u^p\}$ and $\{t^p\}$ are obtained using Eqs. (6.66), (6.67), and (6.69). It is noteworthy that for this case the matrices $\{u^p\}$ and $\{t^p\}$ are direct nodal evaluations at boundary nodes. It is clear that the particular integral approach provides a powerful tool to account for body forces in boundary elements while retaining the boundary-only character of the method. The particular integral approach has been extended to solve a wide variety of problems that result in volume integrals. These include free vibration analysis (Wang and Banerjee, 1990; Wilson et al., 1990), thermoelastic analysis (Saigal et al., 1990b), and elastoplasticity, among others.

6.14 EVALUATION OF STRESSES AND STRAINS

The stresses and strains within a body are often the engineering quantities of interest in an analysis. A procedure will now be formulated for the boundary element method that yields these quantities given the displacements and tractions on the boundary of the object. A direct procedure may be to substitute the relation for displacement $u_i(P)$ given in Eq. (6.47) into the strain-displacement relation of Eq. (6.10) to obtain the strains at point P. Substituting these strains into the stress-strain relations in Eq. (6.11) can next yield the stresses at point P. It is noted, however, that for determining boundary stresses the integral relation in Eq. (6.47) contains U_{ij}^* which for two-dimensional analysis leads to a strongly singular integral due to the presence of $1/r$ terms in the integrand. A differentiation of Eq. (6.47) as required by the strain–displacement relation in Eq. (6.10) will yield $1/r^2$ terms that lead to a hypersingular integral. Several techniques have been developed recently for the treatment of hypersingular integrals, and much research in this direction is currently ongoing.

A procedure is presented here that avoids the evaluation of hypersingular integrals and allows the computation of boundary stresses directly from boundary displacements and tractions. The displacements and tractions at nodal locations on the boundary are known through the solution of Eq. (6.54). These quantities at any other location on the boundary can easily be obtained using shape functions as given in Eqs. (6.37a) and (6.37b). A local coordinate system is now constructed at the point P of interest. The coordinate axes for this local coordinate system lie along the tangents and the normal to the surface at that point, as shown in Fig. 6.10. The displacements and tractions at point P with respect to the local coordinate system are now obtained as

$$\{u'\} = [R][N]\{u\} \tag{6.70}$$

$$\{t'\} = [R][N]\{t\} \tag{6.71}$$

where $\{u\}$ and $\{t\}$ are the nodal displacements vectors, $[N]$ are the shape functions matrices to obtain displacements and tractions at the point P of interest, and $[R]$ is the coordinate transformation matrix. For two-dimensional analysis, the matrix $[R]$ is given as

$$[R] = \begin{bmatrix} \cos\theta & \sin\theta \\ -\sin\theta & \cos\theta \end{bmatrix} \tag{6.72}$$

where θ is the angle between the coordinate axes x_1 and x_1' (Fig. 6.10b).

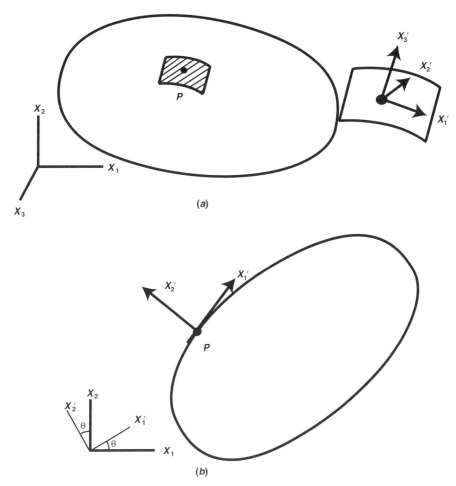

Figure 6.10 Local coordinate system at a point on the boundary. (a) Three-dimensional object. (b) Two-dimensional object.

It is noted that several stress components in the local coordinate systems are already available as

$$\sigma'_{13} = \sigma'_{31} = t'_1$$

$$\sigma'_{23} = \sigma'_{32} = t'_2 \qquad (6.73)$$

$$\sigma'_{33} = t'_3$$

where the primed tractions were calculated in Eq. (6.71). The strains in the local coordinate system are computed as

$$\varepsilon'_{ij} = \tfrac{1}{2}(u'_{j,i} + u'_{i,j}) \tag{6.74}$$

A typical derivative, $u'_{i,j}$, in Eq. (6.74) needs to be computed to obtain the strain ε'_{ij}. Using the chain rule of differentiation gives us

$$\frac{\partial u'_i}{\partial x'_j} = \frac{\partial u'_i}{\partial \xi} \frac{\partial \xi}{\partial x'_j} = \frac{\partial u'_i}{\partial \xi} J^{-1} \tag{6.75}$$

where $J = \partial x'_j / \partial \xi$ is the Jacobian at point P. The term $\partial u'_i / \partial \xi$ on the right-hand side of Eq. (6.75) is computed using shape functions $[N]$. The displacements at point P in the local coordinate system are expressed as

$$u'_i = \sum_{k=1}^{\text{NNODE}} N_k u'^{(k)}_i \tag{6.76}$$

where NNODE represents the number of nodes of the element within which the point P lies and $u'^{(k)}_i$ represents the nodal displacements of that element in the local coordinate system. The derivative $\partial u'_i / \partial \xi$ is then obtained as

$$\frac{\partial u'_i}{\partial \xi} = \sum_{k=1}^{\text{NNODE}} \frac{\partial N_k}{\partial \xi} u'^{(k)}_i \tag{6.77}$$

The Jacobian, J, is similarly written as

$$\frac{\partial x'_i}{\partial \xi} = \sum_{k=1}^{\text{NNODE}} \frac{\partial N_k}{\partial \xi} x'^{(k)}_i \tag{6.78}$$

where $x'^{(k)}_j$ are the nodal coordinates of node k expressed in the local coordinate system.

The derivatives $\partial N_k / \partial \xi$ are obtained through differentiation of relations in Eq. (6.36). Using Eqs. (6.77) and (6.78), the strain components ε'_{ij} can now be obtained easily. In the final step, these strains are substituted in the stress–strain relations given in Eq. (6.11) to obtain the remaining stress components in the local coordinate system as

$$\sigma'_{12} = \sigma'_{21} = 2\mu\varepsilon'_{12} \tag{6.79}$$

$$\sigma'_{11} = \frac{1}{1 - \nu} \left[\nu\sigma'_{33} + 2\mu(\varepsilon'_{11} + \varepsilon'_{22}) \right]$$

where ν is Poisson's ratio. The stress components σ'_{ij} in local coordinate system are all available from Eqs. (6.73) and (6.79). The stress components σ_{ij} may now be obtained from σ'_{ij} using a tensor coordinate transformation (Chung, 1988).

6.15   CORNER PROBLEM IN THE BOUNDARY ELEMENT METHOD

The presence of sharp corners in the boundary of an object causes difficulties in the boundary element analysis approach. This is primarily because tractions at such sharp corners are not continuous. In two dimensions, for example, two additional traction components exist at a node located at a sharp corner compared to when it is located at a smooth corner, as shown in Fig. 6.11. The problem may first be realized for the case when displacements $\{u\}$ are prescribed everywhere on the boundary. For this case, from Eq. (6.48), it is seen that we will need to solve the matrix equation

$$[G]\{t\} = \{f_u\} \tag{6.80}$$

where $\{f_u\} = [F]\{u\}$ is the force term on the right-hand side due to the prescribed displacements $\{u\}$. The matrix $[G]$, however, is rectangular, as explained earlier, and a solution to Eq. (6.80) may or may not be obtained. In general, for two-dimensional analysis, only two unknowns may exist at any node location. These could be displacements, tractions, or a combination of displacement and traction. If there are, indeed, only two unknowns at a node, no problems arise and the boundary element solution is easily obtained. Although this is always true at a smooth boundary, more than two unknowns may exist at a sharp corner. At such corners, six variables—two displacement and four traction components—exist. If four of these are prescribed through boundary conditions, leaving two of them unknown, no problems exist. If, on the other hand, less than four of these variables are prescribed, the boundary element method fails to yield a solution. Consider, for ex-

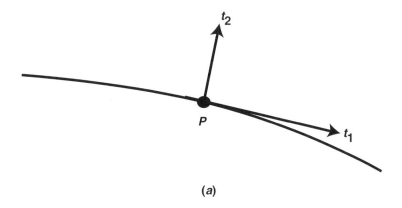

(*a*)

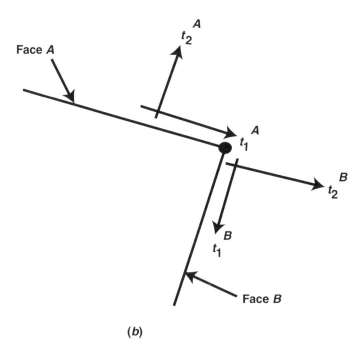

(*b*)

Figure 6.11   Traction components at a node on a smooth boundary and at a sharp corner. (*a*) Smooth boundary. (*b*) Sharp corner.

ample, the sharp corner shown in Fig. 6.12. The displacements on both faces meeting at the corner node are prescribed. Thus, the corner node in this situation will have four unknown variables, two traction components from each of the faces meeting at the sharp corner. This will result in more unknowns than the number of equations available from the overall boundary element system. A number of remedies to the corner problem have been proposed. Some of them are ad hoc in nature, while others are based on advanced concepts in elasticity. A few of these remedies are described here.

Corner Smoothing   The sharp corner is eliminated by smoothing the corner. The resulting geometry is smooth, thus no problem exists. This procedure may be unreliable for cases where the presence of corner is important to the overall response of the body.

Double-Node Representation   Two nodes, instead of one, are provided at the sharp corner. Thus, for the sharp corner shown in Fig. 6.12, element m will have a node C_1 at the corner C while the element n will have a node C_2 at the same corner. The two nodes C_1 and C_2 are carried in the assembled boundary element equations (6.48) as two separate nodes; thus each can have its own set of tractions. It is noted that each row in the matrices $[F]$ and $[G]$ of Eq. (6.48) corresponds to evaluation of the boundary integrals in Eq. (6.47) with respect to a node on the boundary being a load point. Thus, the two rows corre-

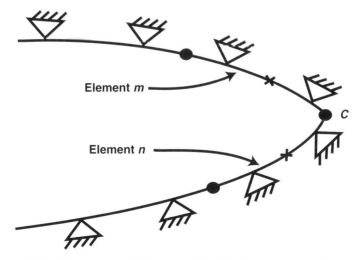

Figure 6.12   Sharp corner with "corner problem" in boundary element approach.

sponding to nodes C_1 and C_2 lying at the same location will be identical, leading to a singular $[F]$ matrix. To avoid this situation, the two nodes are typically separated by a small distance ε. This distance must be carefully prescribed. A small value may lead to a singular $[F]$ matrix, while a large value may alter the physical geometry of the object.

Discontinuous Boundary Elements   Special boundary elements, known as *discontinuous boundary elements,* may be formulated for use in the vicinity of the corner point. These special elements have one or both of their end nodes not lie at the ends of the elements but at a small distance inside. This requires the use of modified shape functions to account for the new location of the end node. Since the geometry of the boundary is also interpolated using the same end nodes and the same shape functions, the use of discontinuous boundary elements may lead to geometries with gaps and overlaps at the end. It is also possible to formulate a boundary element using a different set of shape functions and end nodes for interpolating the geometry and the solution variables (tractions and displacements). Discontinuous boundary elements were popular in the earlier implementations of boundary elements but have largely been abandoned, due to their poor performance.

Finite Element at Corner   The corner problem in boundary elements may also be resolved by providing a finite element at the corner as shown in Fig. 6.13. This approach is not very popular and may lead

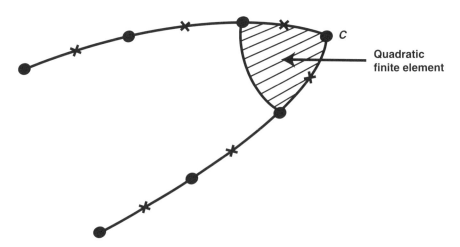

Figure 6.13   Finite element at a boundary corner.

to ambiguity in the choice of normal required in formulation of the finite element.

Several other approaches have been proposed in the boundary element literature. Typically, these approaches provide two additional equations based on advanced elasticity concepts (see, e.g., Chaudonneret, 1978). These equations augment the boundary element equations of Eq. (6.48) to account for the special behavior at the corner. The inclusion of these additional equations complicates the assembly procedure and is not very desirable. Most commercial implementations use the double-node representation of the sharp corner to treat the corner problem.

6.16 CLOSING REMARKS

The boundary element method has emerged as a powerful numerical analysis tool in the last three decades. Many engineers erroneously believe the method to be applicable only to linear problems, perhaps due to the early difficulties with nonlinear boundary element formulations. The method has been applied for the solution of a wide variety of linear and nonlinear problems in materials (Mukherjee, 1982) and displacements (Banerjee and Cathie, 1980; Banerjee and Raveendra, 1986; Banerjee et al., 1989). Initial formulations of nonlinear problems required the discretization of the domain of the body into integration cells. Techniques such as the particular integral approach and the multiple reciprocity method have been extended to derive boundary-only integral relations for nonlinear problems. The boundary element method has had special success in far-field problems of geomechanics (Brebbia, 1988b), acoustics (Banerjee et al., 1988; Bernhard and Keltie, 1989), and electromagnetics (Brebbia, 1988a) due to its ability to account exactly for the far-field effects (Kaljevic et al, 1992). The method has also been applied to the analysis of inclusion problems (Banerjee and Henry, 1991, 1992), contact problems (Simunovic and Saigal, 1995), shape sensitivity analysis (Kane and Saigal, 1988; Saigal, 1989; Saigal et al., 1990a, Saigal and Chandra, 1991), thermoelasticity (Rizzo and Shippy, 1977; Saigal et al., 1990b), stochastic analysis (Ettouney et al., 1989; Kaljevic and Saigal, 1993, 1995b), fluid flow (Brebbia, 1989a), and the highly nonlinear problems of metalworking (Chandra and Mukherjee, 1984a,b), including sheet metal forming, extrusion

(Chandra and Saigal, 1991), and so on. While the boundary element method certainly does not enjoy the popularity of the finite element method with the analysts, it has been shown to be a powerful method in its own right. Application of the boundary element method is largely restricted to bulky objects. Its application to thin geometries, especially in parts where thin geometries exist next to bulky parts, may lead to ill-conditioned matrices. This can be understood by looking at any individual row of the matrix $[F]$ in Eq. (6.48) to correspond to a load point P on the boundary. For thin geometries, two adjacent nodes may lie approximately at the same location (especially in comparison with nodes for a neighboring bulky object). These two nodes will then yield nearly identical rows leading to a singular $[F]$ matrix. The problem may be resolved by providing very fine mesh in the bulky region, comparable in density to the mesh of the thin geometry. Another approach to avoid this problem is through the use of multiple zones (Kane et al., 1990), as shown in Fig. 6.14. Separate boundary element relations are written for zones *ABF*, *BCEF*, and *CDE*, respectively. The overall boundary element matrices are assembled by considering continuity of displacements and tractions at interface surfaces between two zones. In such a procedure, the equations from a zone representing a thin geometry may be scaled appropriately to avoid numerical complications. An additional advantage of the zoning procedure lies in the fact that it imparts a sparse, banded structure to the overall boundary element matrices.

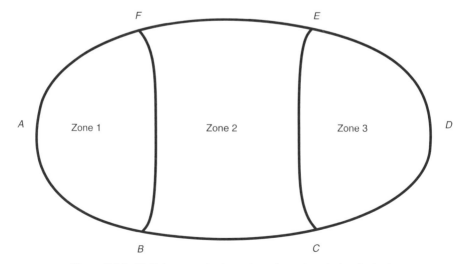

Figure 6.14 Multiple zones for boundary element analysis of a body.

The absence of symmetry in the boundary element matrices has also been a cause for the boundary element method not attaining the popularity that was initially anticipated. This should cease to be a limitation, as the numerical analysis community continues to develop ever-robust algorithms for the solution of nonsymmetric systems of equations. Also, a number of symmetric boundary element formulations have lately been developed. Although these new formulations are not widely used, they have the potential to further enhance the attractiveness of the boundary element method.

REFERENCES

Ahmad, S., and Banerjee, P. K. 1986. Free-vibration analysis by BEM using particular integrals, *ASCE J. Eng. Mech.,* **112,** 682–695.

Balakrishna, C., Gray, L. J., and Kane, J. H. 1994. Efficient analytical integration of symmetric Galerkin boundary integrals over curved elements: elasticity formulation, *Comput. Methods Appl. Mech. Eng.,* **117,** 157–179.

Banerjee, P. K. 1994. *Boundary Element Methods in Engineering,* McGraw-Hill, London.

Banerjee, P. K., and Cathie, D. N. 1980. A direct formulation and numerical elastoplasticity, *Int. J. Mech. Sci.,* **22,** 233–245.

Banerjee, P. K., and Henry, D. 1991. Analysis of three-dimensional solids with holes by BEM, *Int. J. Numer. Methods Eng.,* **31,** 369–384.

Banerjee, P. K., and Henry, D. 1992. Analysis of 3-dimensional solids with inclusions by BEM, *Int. J. Solids Struct.,* **29,** 2423–2440.

Banerjee, P. K., and Raveendra, S. T. 1986. Advanced boundary element method for two- and three-dimensional problems of elastoplasticity, *Int. J. Numer. Methods in Eng.,* **23,** 985–1002.

Banerjee, P. K., and Wilson, R. B. (eds.) 1989. *Industrial Applications of Boundary Element Methods,* Elsevier Applied Science, New York.

Banerjee, P. K., Ahmad, S., and Wang, H. C. 1988. A new BEM formulation for the acoustic eigen frequency analysis, *Int. J. Numer. Methods Eng.* **26,** 1299–1309.

Banerjee, P. K., Henry, D., and Raveendra, S. T. 1989. Advanced BEM for inelastic analysis of solids, *Int. J. Mech. Sci.,* **31,** 309–322.

Bernhard, R. J., and Keltie, R. F. (eds.) 1989. *Numerical Techniques in Acoustic Radiation,* ASME Press, New York.

Boresi, A. P., and Chong, K. P. 2000. *Elasticity in Engineering Mechanics,* 2nd ed., Wiley, New York.

Boyce, W. E., and DiPrima, R. C. 1977. *Elementary Differential Equations and Boundary Value Problems,* 3rd ed., Wiley, New York.

Brebbia, C. A. (ed.) 1988a. *Boundary Elements X,* Vol. 2, *Heat Transfer, Fluid Flow and Electrical Applications,* Computational Mechanics Publications, Ashurst, Southampton, England.

Brebbia, C. A. (ed.) 1988b. *Boundary Elements X,* Vol. 4., *Geomechanics, Wave Propagation and Vibrations,* Computational Mechanics Publications, Ashurst, Southampton, England.

Chandra, A., and Mukherjee, S. 1984a. A finite element analysis of metal forming problems with an elastic-viscoplastic material model, *Int. J. Numer. Methods Eng.,* **20,** 1613–1628.

Chandra, A., and Mukherjee, S. 1984b. A finite element analysis of metal forming processes with thermomechanical coupling, *Int. J. Mech. Sci.,* **26,** 661–676.

Chandra, A., and Saigal, S. 1991. A boundary element analysis of the axisymmetric extrusion processes, *Int. J. Solids Struct,* **26,** 1–13.

Chaudonneret, M. (1978). On the discontinuity of the stress vector in the boundary integral equation method for elastic analysis, in *Recent Advances in Boundary Element Methods,* Brebbia, C. A. (ed.), TA347.B69.R42, pp. 185–194.

Chung, T. J. 1988. *Continuum Mechanics,* Prentice Hall, Upper Saddle River, NJ.

Churchill, R. V., and Brown, J. W. 1978. *Fourier Series and Boundary Value Problems,* McGraw-Hill, New York.

Ettouney, M., Benaroya, H., and Wright, H. 1989. Boundary element methods in probabilisitic structural analysis, *Appl. Math. Model.,* **13,** 432–441.

Gray, L. J., and Paulino, G. H. 1997. Symmetric Galerkin boundary integral formulation for interface and multi-zone problems, *Int. J. Numer. Methods Eng.,* **16,** 3085–3101.

Hartmann, F., Katz, C., and Protopsaltis, B. 1985. Boundary elements and symmetry, *Ing. Arch.,* **55,** 440–449.

Henry, D., and Banerjee, P. K. 1988a. A new boundary element formulation for two and three-dimensional problems of thermoelasticity using particular integrals, *Int. J. Numer. Methods Eng.,* **26,** 2061–2077.

Henry, D., and Banerjee, P. K. 1988b. A new boundary element formulation for two and three-dimensional problems of elastoplasticity using particular integrals, *Int. J. Numer. Methods Eng.,* **26,** 2079–2096.

Henry, D. P., Jr., Pape, D. A., and Banerjee, P. K. 1987. A new axisymmetric BEM formulation for body forces using particular integrals, *ASCE J. Eng. Mech.,* **113,** 671–688.

Kaljevic, I., and Saigal, S. 1993. Stochastic boundary elements in elastostatics, *Comput. Methods Appl. Mech. Eng.,* **109,** 259–280.

Kaljevic, I., and Saigal, S. 1995a. Analysis of symmetric domains in advanced applications with the boundary element method, *Int. J. Numer. Methods Eng.,* **38,** 2373–2388.

Kaljevic, I., and Saigal, S. 1995b. Stochastic boundary elements for two-dimensional potential flow in homogeneous domains, *Int. J. Solids Struct.,* **32,** 1873–1892.

Kaljevic, I., Saigal, S., and Ali, A. 1992. An infinite boundary element formulation for three-dimensional potential problems, *Int. J. Numer. Methods Eng.,* **35,** 2079–2100.

Kane, J. H., and Saigal, S. 1988. Design sensitivity analysis of solids using BEM, *ASCE J. Eng. Mech.,* **114,** 1703–1722.

Kane, J. H., Kumar, B. L. K., and Saigal, S. 1990. An arbitrary condensing, noncondensing solution strategy for large scale, multi-zone boundary element analysis, *Comput. Methods Appl. Mech. Eng.,* **79**(2), 219–244.

Layton, J. B., Ganguly, S., Balakrishna, C., and Kane, J. H. 1997. A symmetric Galerkin multi-zone boundary element formulation, *Int. J. Numer. Methods Eng.,* **16,** 2913–2932.

Love, A. E. H. 1944. *A Treatise on the Mathematical Theory of Elasticity,* Dover Publications, Mineola, NY.

Maier, G., Novati, G., and Sirtori, S. 1990. On symmetrization in boundary element elastic and elastoplastic analysis, in *Discretization Methods in Structural Mechanics,* Kuhn, G., and Mang, H. (eds.), Springer-Verlag, Berlin, pp. 191–200.

Mukherjee, S. 1982. *Boundary Element Methods in Creep and Fracture,* Elsevier Applied Science, Barking, Essex, England.

Neves, A. C., and Brebbia, C. A., 1991. Multiple reciprocity boundary element method in elasticity: a new approach for transforming domain integrals to the boundary, *Int. J. Numer. Methods Eng.,* **31**(4), 709–727.

Niku, S. M., and Brebbia, C. A. 1988. Dual reciprocity boundary element formulation for potential problems with arbitrarily distributed sources, *Engineering Analysis,* **5,** 46–48.

Norris, C. II., Wilbur, J. B., and Utku, S. 1976. *Elementary Structural Analysis,* 3rd ed., McGraw-Hill, New York.

Nowak, A. J., and Neves, A. C. (eds.) 1994. *The Multiple Reciprocity Boundary Element Method,* Wessex Institute of Technology, Ashurst, Southampton, England.

Pape, D. A., and Banerjee, P. K. 1987. Treatment of body forces in 2-D elastostatic BEM using particular integrals, *ASME J. Appl. Mech.,* **54,** 866–871.

Partridge, P. W., and Brebbia, C. A. 1990. Computer implementation of the BEM dual reciprocity method for the solution of general field equations, *Commun. Appl. Numer. Methods,* **6,** 83–92.

Partridge, P. W., Brebbia, C. A., and Wrobel, L. C. 1991. *The Dual Reciprocity Boundary Element Method,* Wessex Institute of Technology, Ashurst, Southampton, England.

Rizzo, F. J., and Shippy, D. J. 1977. An advanced boundary integral equation method for three-dimensional thermoelasticity, *Int. J. Numer. Methods Eng.,* **11,** 1753–1768.

Saigal, S. 1989. Treatment of body forces in axisymmetric boundary element design sensitivity formulation, *Int. J. Solids Struct.,* **25,** 947–959.

Saigal, S., and Chandra, A. 1991. Shape sensitivities and optimal configurations for heat diffusion problems: a BEM approach, *ASME J. Heat Transfer,* **113,** 287–295.

Saigal, S., and Kane, J. H. 1990. Boundary element shape optimization system for aircraft structural components, *AIAA J.,* **28,** 1203–1204.

Saigal, S., Aithal, R., and Dyka, C. T. 1990a. Boundary element design sensitivity analysis of symmetric bodies, *AIAA J.,* **28**(1), 180–183.

Saigal, S., Gupta, A., and Cheng, J. 1990b. Stepwise linear regression particular integrals for uncoupled thermoelasticity with boundary elements, *Int. J. Solids Struct.* **26,** 471–482.

Schey, H. M. 1973. *Div, Curl, and All That,* W.W. Norton, New York.

Simunovic, S., and Saigal, S. 1995. A linear programming formulation for incremental contact analysis, *Int. J. Numer. Methods Eng.,* **38,** 2703–2725.

Sladek, V., and Sladek, J. 1996. Multiple reciprocity method in BEM formulations for solution of plate bending problems, *Eng. Anal. Boundary Elements,* **17,** 161–173.

Sokolnikoff, I. 1956. *Mathematical Theory of Elasticity,* McGraw-Hill, New York.

Somigliana, C. 1886. Sópra l'equilìbriodi un còrpo elàstico isotropo, *Nuovo Cimento,* pp. 17–19.

Stewart, J. 1999. *Calculus Early Transcendentals,* 4th ed., Brooks/Cole Pacific Grove, CA.

Wang, H. C., and Banerjee, P. K. 1990. Generalized axisymmetric free-vibration analysis by BEM, *Int. J. Numer. Methods Eng.,* **29,** 985–1001.

Wilson, R. B., Miller, N. M., and Banerjee, P. K. 1990. Calculations of natural frequencies and mode shapes for three-dimensional structures by BEM, *Int. J. Numer. Methods Eng.,* **29,** 1737–1757.

Wrobel, L. C., and Brebbia, C. A. 1987. Dual reciprocity boundary element formulation for nonlinear diffusion problems, *Comput. Methods Appl. Mech. Eng.,* **65**(2), 147–164.

Zeng, X., and Saigal, S. 1991. A Fourier expansion based particular integral approach for non-boundary loadings in boundary element method, *Commun. Appl. Numer. Methods,* **7,** 453–464.

7

Meshless Methods of Analysis

7.1 INTRODUCTION

A series of meshless methods for analysis have been introduced in the literature, mostly in the late 1990s. The development of these methods may have been motivated by the shortcomings arising in the difficulty of use of the finite element methods by analysts. Some of the shortcomings were outlined in Section 6.1. Methods in the category of meshless methods include smooth particle hydrodynamics (SPH) (Swegle et al., 1994), the element-free Galerkin (EFG) method (Belytschko et al., 1994a; Lu et al., 1994), reproducing kernel particle methods (RKPM) (Liu, 1995; Liu and Chen, 1995; Liu et al., 1996), partition of unity methods (PUM) (Melenk and Babuska, 1996), h-p clouds (Duarte and Oden, 1996), the meshless local Petrov–Galerkin (MLPG) method (Atluri and Zhu, 1998, 2000a; Atluri and Shen, 2002), the natural element method (Sukumar et al., 1998), the natural neighbor Galerkin methods (Sukumar et al., 2001), and the method of finite spheres (De and Bathe, 2000, 2001). Except for the SPH methods, these methods are all based, in principle, on the finite element method. Several additional meshless methods based on the boundary element method have also been proposed. These include boundary contour methods (BCMs) (Nagarajan et al., 1994, 1996, Mukherjee et al., 1997), the boundary node method (BNM) (Mukherjee and Mukherjee, 1997a; Chati et al., 1999; Chati and Mukherjee, 2000), and the local boundary integral equation (LBIE) method (Atluri et al., 2000). Most of these methods are of recent origin and are still undergoing development. Some applications of meshless methods in modeling life-cycle engineering appeared in a recent book by Chong et al., 2002. A popular meshless method is the EFG method. It has been developed largely in the context of elastostatics. Chapters 4 and 6 have both been presented for use in elastostatics analysis. Continuing in that vein, the EFG method is also presented in this chapter for linear elastic analysis of solid bodies.

The step of meshing is a cumbersome, time-consuming step, especially for arbitrarily shaped three-dimensional objects. Meshless methods have been developed with the intent to eliminate the need to define a mesh for the object being analyzed. None of the methods developed to date, however, are truly meshless, as they still have the need to define a background grid. A significant shortcoming of these methods is that they involve a heavy computational effort compared to the finite element method. It is to be understood that these methods are all of recent origin and have not enjoyed the growth that the finite element method has seen over the last five decades. Despite their known shortcomings, these methods hold significant promise, as developments for their refinement and better understanding continue. The EFG method is covered first in detail in this chapter, followed by an introduction to some of the other meshless analysis methods.

7.2 EQUATIONS OF ELASTICITY

The analysis of linear elastic behavior of solid bodies (Section 6.3) is considered to illustrate the EFG formulation. The stress components σ_{ij} must satisfy the differential equations of equilibrium in a domain, Ω, of the body under consideration

$$\sigma_{ij,j} + F_i = 0 \tag{7.1}$$

where F_i are the components of the body force acting on the object. The field equations above are to be solved given certain boundary conditions as

$$u_i = \bar{u}_i \qquad \text{on } \Gamma_u$$

$$t_i = \bar{t}_i \qquad \text{on } \Gamma_t \tag{7.2}$$

where the tractions on a surface with normal n_j are given as $t_i = \sigma_{ij}n_j$, and $\bar{u}_i$ and $\bar{t}_i$ are the prescribed values of displacements and tractions on boundaries Γ_u and Γ_t, respectively. $\Gamma_u \cup \Gamma_t = \Gamma$ is the boundary of the subject domain Ω. The stress components σ_{ij} in Eq. (7.1) are symmetric (i.e., $\sigma_{ij} = \sigma_{ji}$). The kinematic equations, relating the strain components ε_{ij} to the displacements u_i are

$$\varepsilon_{ij} = \tfrac{1}{2}(u_{i,j} + u_{j,i}) \tag{7.3}$$

where a comma (,) denotes differentiation; thus, $u_{i,j} = \partial u_i / \partial x_j$. The constitutive relations for a linear elastic material are given as

$$\sigma_{ij} = \lambda \varepsilon_{kk} \delta_{ij} + 2\mu \varepsilon_{ij} = E_{ijkl} \varepsilon_{kl} \tag{7.4}$$

where λ and μ are Lamé's constants and E_{ijkl} is the constitutive tensor. Equations (7. 1) to (7.4) summarize the equations of elastostatics for a linear, isotropic, homogeneous elastic object.

7.3 WEAK FORMS OF THE GOVERNING EQUATIONS

The first law of thermodynamics states that for a body in equilibrium and subjected to arbitrary virtual displacement δv_i, the variation in work of the external forces δW is equal to the variation of internal energy δU. These quantities are written as

$$\delta W - \delta U = \delta W_S + \delta W_B - \delta U = 0 \tag{7.5}$$

$$\delta W_S = \int_{\Gamma_t} \bar{t}_i \, \delta v_i \, d\Gamma \tag{7.6}$$

$$\delta W_b = \int_{\Omega} F_i \, \delta v_i \, d\Omega \tag{7.7}$$

$$\delta U = \int_{\Omega} \sigma_{ij} \, \delta \varepsilon_{ij} \, d\Omega \tag{7.8}$$

where $\delta \varepsilon_{ij} = \tfrac{1}{2}(\delta v_{i,j} + \delta v_{j,i})$ is the strain due to the virtual displacement field δv_i. Since $\sigma_{ij} = \sigma_{ji}$, the product $\sigma_{ij} \, \delta \varepsilon_{ij}$ may be expressed using Eq. (7.3) as $\sigma_{ij} \, \delta v_{ij}$.

In the context of computational mechanics, the displacement field u_i that leads to the stress field σ_{ij} is referred to as the *trial function,* and the virtual displacement field δv_i is referred to as the *test function.*

Both sides of the equilibrium relation in Eq. (7.1) are multiplied by the test function and integrated over the domain Ω to give

$$\int_\Omega (\sigma_{ij,j} + F_i)\, \delta v_i\, d\Omega = 0 \qquad (7.9)$$

Using the result $\sigma_{ij,j}\, \delta v_i = (\sigma_{ij}\, \delta v_i)_{,j} - \sigma_{ij} \delta v_{i,j}$ from differentiation by parts, we get

$$\int_\Omega (\sigma_{ij}\, \delta v_i)_{,j}\, d\Omega - \int_\Omega \sigma_{ij}\, \delta v_{i,j}\, d\Omega + \int_\Omega F_i\, \delta v_i\, d\Omega = 0 \qquad (7.10)$$

The divergence theorem of calculus (Schey, 1973) for a vector field with components R_j is written as

$$\int_\Omega R_{j,k}\, d\Omega = \int_\Gamma R_j n_k\, d\Gamma \qquad (7.11)$$

where Γ is the bounding surface for the volume Ω and n_k is the outward normal to the surface $d\Gamma$. Applying divergence theorem to the first term in Eq. (7.10) and noting that $\sigma_{ij} n_j = t_i$, we obtain

$$\int_\Gamma t_i\, \delta v_i\, d\Gamma - \int_\Omega \sigma_{ij} \delta v_{i,j}\, d\Omega + \int_\Omega F_i\, \delta v_i\, d\Omega = 0 \qquad (7.12)$$

The test functions are specified such that they satisfy the essential boundary conditions. Thus, $\delta v_i = 0$ everywhere on Γ except on the portion Γ_t where $t_i = \bar{t}_i$ from Eq. (7.2), and Eq. (7.12) may be written to obtain a weak form of the governing equations as:

$$\mathrm{W}_1: \qquad \int_{\Gamma_t} t_i\, \delta v_i\, d\Gamma + \int_\Omega F_i\, \delta v_i\, d\Omega - \int_\Omega \sigma_{ij}\, \delta v_{i,j}\, d\Omega = 0 \qquad (7.13)$$

It is noted that Eq. (7.13) is the same relation as the virtual work relation in Eqs. (7.5) to (7.8). Equation (7.13) is adequate when the test function u_i satisfies the essential boundary conditions of Eq. (7.2) i.e., $u_i = \bar{u}_i$ on Γ_u. Alternate forms are resorted to when these conditions are not met. A Lagrange multiplier form (see Belytschko et al., 1994a) may be written by appropriately augmenting Eq. (7.13) to give the weak form

$$W_2: \quad \int_\Omega \sigma_{ij} \, \delta v_{i,j} \, d\Omega - \int_\Omega F_i \, \delta v_i \, d\Omega - \int_{\Gamma_t} \bar{t}_i \, \delta v_i \, d\Gamma$$

$$- \int_u (u_i - \bar{u}_i) \, \delta \lambda_i \, d\Gamma - \int_{\Gamma_u} \lambda_i \, \delta v_i \, d\Gamma = 0 \quad (7.14)$$

where λ_i and $\delta \lambda_i$ are the Lagrange multiplier and its variation, respectively. From heuristic arguments, Lu et al. (1994) concluded that the physical meaning of the Lagrange multipliers λ_i is the traction t_i along the boundary Γ_u. Thus, replacing λ_i by t_i in Eq. (7.14), an additional weak form is obtained as

$$W_3: \quad \int_\Omega \sigma_{ij} \, \delta v_{i,j} \, d\Omega - \int_\Omega F_i \, \delta v_i \, d\Omega - \int_{\Gamma_t} \bar{t}_i \, \delta v_i \, d\Gamma$$

$$- \int_{\Gamma_u} (u_i - \bar{u}_i) \, \delta t_i \, d\Gamma - \int_{\Gamma_u} t_i \, \delta v_i \, d\Gamma = 0 \quad (7.15)$$

The essential boundary conditions can also be accounted for by means of a penalty formulation (Belytschko et al., 1994b; Hegen, 1996; Zhu and Atluri, 1998) to give

$$W_4: \qquad \int_\Omega \sigma_{ij} \, \delta v_{i,j} \, d\Omega - \int_\Omega F_i \, \delta v_i \, d\Omega - \int_{\Gamma_t} \bar{t}_i \, \delta v_i \, d\Gamma$$

$$+ \alpha \int_{\Gamma_u} (u_i - \bar{u}_i) \, \delta v_i \, d\Gamma = 0 \qquad (7.16)$$

where $\alpha \gg 1$ is a penalty parameter to enforce $u_i = \bar{u}_i$ on Γ_u. The choice of the weak form employed in a formulation depends on the strategy adopted therein to satisfy the essential boundary conditions.

7.4 MOVING LEAST SQUARES APPROXIMATIONS

In the element-free Galerkin method, as well as several other meshless methods, the test and trial functions are represented over the domain of the body using moving least squares (MLS) approximants. The details for such approximations were presented in Lancaster and Salkauskas (1980). To define an MLS approximation over domain Ω, a set of nodal points x_i, $i = 1, 2, \ldots, N$ are used. The MLS approximant u^h of the function u at a location x is given by

$$u^h(x) = \sum_{j=1}^{m} p_j(x)a_j(x) = \lfloor p(x) \rfloor \{a(x)\} \qquad (7.17)$$

where

$$\lfloor p(x) \rfloor = \lfloor p_1(x), p_2(x), \ldots, p_m(x) \rfloor \qquad (7.18)$$

$$\{a(x)\}^{\mathrm{T}} = \lfloor a_1(x), a_2(x), \ldots, a_m(x) \rfloor \qquad (7.19)$$

$\{a(x)\}$ is a vector of coefficients that depend on the location x and $\lfloor p(x) \rfloor$ is a vector of monomials called *basis functions*. The monomials are chosen so that they provide a complete basis of order m. For a two-dimensional domain, complete linear and quadratic bases are provided by

Linear: $\qquad \lfloor p(x) \rfloor = \lfloor 1, x, y \rfloor \qquad\qquad m = 3 \qquad (7.20)$

Quadratic: $\qquad \lfloor p(x) \rfloor = \lfloor 1, x, y, x^2, xy, y^2 \rfloor \qquad m = 6 \qquad (7.21)$

The approximation error over the entire domain may be characterized by a weighted discrete L_2 norm defined at the N nodal locations as (Lancaster and Salkauskas, 1980)

$$J(\{a(x)\}) = \sum_{i=1}^{N} \left[u^h(x_i) - u(x_i) \right]^2 w_i(x)$$

$$= \sum_{i=1}^{N} \left[\sum_{s=1}^{m} p_s(x_i)a_s(x) - u_i \right]^2 w_i(x) \qquad (7.22)$$

where $u(x_i)$ are the generalized displacements at nodal locations x_i and w_i are weight functions. Obviously, for $u^h(x)$ to be a good approximation for $u(x)$, the error $J(\{a(x)\})$ must be a minimum. The coefficients $\{a(x)\}$ in Eq. (7.22) are still undetermined. These coefficients need to be selected in such a way that the error $J\{a(x)\}$ is minimized. This leads to the following m conditions corresponding to the m unknown coefficients:

$$\frac{\partial J\{a(x)\}}{\partial a_k(x)} = 0 \qquad k = 1, 2, \ldots, m \qquad (7.23)$$

Substituting for J from Eq. (7.22), we get

$$\frac{\partial J}{\partial a_k} = \sum_{i=1}^{N} 2[u^h(x_i) - u(x_i)] \frac{\partial u^h(x_i)}{\partial a_k} w_i(x) \qquad (7.24)$$

where the derivative

$$\frac{\partial u^h(x_i)}{\partial a_k} = \frac{\partial}{\partial a_k} \left[\sum_{s=1}^{m} p_s(x_i) a_s(x) \right]$$

$$= \sum_{s=1}^{m} p_s(x_i) \frac{\partial a_s(x)}{\partial a_k(x)} \qquad (7.25)$$

Since $\partial a_s(x)/\partial a_k(x) = 1$ when $s = k$ and 0 otherwise, the summation above reduces to $p_k(x_i)$. Thus,

$$\frac{\partial u^h(x_i)}{\partial a_k} = p_k(x_i) \qquad (7.26)$$

Substituting Eq. (7.26) into Eq. (7.24) and plugging the result in Eq. (7.25), upon cancellation of the factor of 2 gives, with Eq. (7.22),

$$\sum_{i=1}^{N} [u^h(x_i) - u(x_i)] p_k(x_i) w_i(x)$$

$$= \sum_{i=1}^{N} \left[\sum_{s=1}^{m} p_s(x_i) a_s(x) - u(x_i) \right] p_k(x_i) w_i(x) = 0 \qquad (7.27)$$

Upon rearranging terms, we obtain

$$\sum_{i=1}^{N} \left[\sum_{s=1}^{m} p_s(x_i) a_s(x) \right] p_k(x_i) w_i(x) = \sum_{i=1}^{N} u(x_i) p_k(x_i) w_i(x) \qquad (7.28)$$

Equation (7.28) represents m equations corresponding to $k = 1, 2, \ldots, m$. To aid the beginner in writing the matrix form of Eq. (7.28), the expanded form of right-hand side of Eq. (7.28) is written as, for $k = 1$,

$$u(x_1)p_1(x_1)w_1(x) + u(x_2)p_1(x_2)w_2(x) + \cdots + u(x_N)p_1(x_N)w_N(x) \quad (7.29)$$

and for $k = 2, \ldots, m$,

$$u(x_1)p_2(x_1)w_1(x) + u(x_2)p_2(x_2)w_2(x) + \cdots + u(x_N)p_2(x_N)w_N(x) \quad (7.30)$$

$$\vdots$$

$$u(x_1)p_m(x_1)w_1(x) + u(x_2)p_m(x_2)w_2(x) + \cdots + u(x_N)p_m(x_N)w_N(x) \quad (7.31)$$

The matrix form of expressions in Eqs. (7.29) to (7.31) can now easily be seen to be

$$
\underbrace{\begin{bmatrix} p_1(x_1) & p_1(x_2) & \cdots & p_1(x_N) \\ p_2(x_1) & p_2(x_2) & \cdots & p_2(x_N) \\ \vdots & & & \\ p_m(x_1) & p_m(x_2) & \cdots & p_m(x_N) \end{bmatrix}}_{m \times N}
$$

$$
\underbrace{\begin{bmatrix} w_1(x) & & & [0] \\ & w_2(x) & & \\ [0] & & \ddots & \\ & & & w_N(x) \end{bmatrix}}_{N \times N} \underbrace{\begin{Bmatrix} u(x_1) \\ u(x_2) \\ \vdots \\ u(x_N) \end{Bmatrix}}_{N \times 1} \quad (7.32)
$$

or in compact matrix notation as

$$\text{RHS} = [P]^{\mathrm{T}}[W]\{u\} \quad (7.33)$$

Aided by the expressions in Eqs. (7.32) and (7.33), Eq. (7.28) can be written in matrix form as

$$[P]^{\mathrm{T}}_{m \times N}[W]_{N \times N}[P]_{N \times m}\{a(x)\}_{m \times 1} = [P]^{\mathrm{T}}_{m \times N}[W]_{N \times N}\{u\}_{N \times 1} \quad (7.34)$$

$$[P]^{\mathrm{T}} = \begin{bmatrix} p_1(x_1) & p_1(x_2) & \cdots & p_1(x_N) \\ p_2(x_1) & p_2(x_2) & \cdots & p_2(x_N) \\ \vdots & & & \\ p_m(x_1) & p_m(x_2) & & p_m(x_N) \end{bmatrix} \quad (7.35)$$

$$[W] = \begin{bmatrix} w_1(x) & \cdots & & [0] \\ \vdots & w_2(x) & & \vdots \\ & & \ddots & \\ [0] & \cdots & & w_N(x) \end{bmatrix} \quad (7.36)$$

$$\{a(x)\} = \begin{Bmatrix} a_1(x) \\ a_2(x) \\ \vdots \\ a_m(x) \end{Bmatrix} \quad (7.37)$$

$$\{u\} = \begin{Bmatrix} u(x_1) \\ u(x_2) \\ \vdots \\ u(x_N) \end{Bmatrix} \quad (7.38)$$

Introducing the short-hand notation

$$[A(x)]_{m \times m} = [P]^{\mathrm{T}}[W][P] \quad (7.39)$$

The unknown coefficients $\{a(x)\}$ are obtained from Eq. (7.34) as

$$\{a(x)\} = [A(x)]^{-1}[P]^{\mathrm{T}}[W]\{u\} \quad (7.40)$$

Substituting Eq. (7.40) into Eq. (7.17), we obtain

$$\{u^h\} = [P]^{\mathrm{T}}[A(x)]^{-1}[P]^{\mathrm{T}}[W]\{u\} \quad (7.41)$$

with

$$\{u^h\} = \begin{Bmatrix} u^h(\underset{\sim}{x}_1) \\ u^h(\underset{\sim}{x}_2) \\ \vdots \\ u^h(\underset{\sim}{x}_N) \end{Bmatrix} \tag{7.42}$$

From Eq. (7.40), we note that

$$a_j(x) = [[A(\underset{\sim}{x})]^{-1}[P]^{\mathrm{T}}[W]]_{j\mathrm{th\,row}}\{u\}$$

$$= \sum_{i=1}^{N} [[A(\underset{\sim}{x})]^{-1}[P]^{\mathrm{T}}[W]]_{ji} u_i \tag{7.43}$$

where $u_i = u(\underset{\sim}{x}_i)$. Substituting in Eq. (7.17), we obtain

$$u^h(\underset{\sim}{x}) = \sum_{j=1}^{m} p_j(\underset{\sim}{x}) \sum_{i=1}^{N} [[A(\underset{\sim}{x})]^{-1}[P]^{\mathrm{T}}[W]]_{ji} u_i \tag{7.44}$$

The order of the two summations in Eq. (7.44) can be changed to obtain

$$u^h(\underset{\sim}{x}) = \sum_{i=1}^{N} u_i \sum_{j=1}^{m} p_j(x)[[A(x)]^{-1}[P]^{\mathrm{T}}[W]]_{ji}$$

$$= \sum_{i=1}^{N} u_i \phi_i(\underset{\sim}{x}) \tag{7.45}$$

with

$$\phi_i(\underset{\sim}{x}) = \sum_{j=1}^{m} p_j(\underset{\sim}{x})[[A(\underset{\sim}{x})]^{-1}[P]^{\mathrm{T}}[W]]_{ji} \tag{7.46}$$

where $\phi_i(x)$ are the shape functions obtained from MLS approximation. This approximation is possible only when a unique solution exists for the system of equations (7.40). The matrix $[A(\underset{\sim}{x})]$ of Eq. (7.39) must be invertible. This is the case if and only if the rank of matrix $[P]$ equals m, the number of basis functions employed.

7.5 CHARACTERISTICS OF MLS APPROXIMATION

The expression in Eq. (7.45) provides an approximation for the displacement field using the shape functions $\phi_i(x)$ defined in Eq. (7.46). Even when the basis functions $p_j(x)$ in Eq. (7.46), which also appear in the matrix $[P]$ in that equation, are polynomials, the MLS approximation in Eq. (7.45) is not a polynomial. It can be shown (see Nayroles et al., 1992) that if $u(x)$ is a polynomial, it is reproduced exactly by the approximation of $u^h(x)$ in Eq. (7.45). Thus, for any polynomial function $g(x)$,

$$\sum_{i=1}^{N} \phi_i(x)g(x_i) = g(x) \tag{7.47}$$

For $g(x) = 1$, we obtain

$$\sum_{i=1}^{N} \phi_i(x) = 1 \tag{7.48}$$

Equation (7.48), called the *partition of unity* (Melenk and Babuska, 1996), reproduces a rigid-body field exactly. The necessary condition for convergence in a Galerkin formulation is the completeness of the interpolations for trial and test functions. The fundamental requirement for completeness is the ability of the functions to represent a constant-valued field as is satisfied by the MLS shape functions in Eq. (7.48).

Unlike the shape functions employed in the finite element method, the EFG shape functions do not satisfy the *Kronecker delta criterion*, also termed the *selectivity property*, so that $\phi_i(x_j) \neq \delta_{ij}$. This property of the EFG shape functions causes difficulty in the direct imposition of the essential boundary conditions (i.e., $u_i = \bar{u}_i$ on Γ_u; see Section 7.9).

The smoothness of the EFG shape functions ϕ_i is determined by the smoothness of the basis functions p_j and the weight functions w_i. Let C^k (Ω) denote a set of functions that have continuous derivatives up to order k on Ω. If $p_j(x) \in C^s$ (Ω) and $w_j(x) \in C^r(\Omega)$, then $\phi_i(x) \in C^t(\Omega)$, where $t = \min(s, r)$.

As seen from the expression for $\phi_i(x)$ in Eq. (7.46), the inversion of matrix $[A]$ must be performed at each nodal location to obtain the shape function for that location. This is a computationally burdensome step in the EFG formulation. An alternative procedure that avoids the inversion step is based on the orthogonalization of the basis functions with respect to the values of weight functions in x (Lu et al., 1994). Although the orthogonalization procedure is preferred from the point of view of accuracy, it involves the same order of costs as the matrix inversion.

To obtain the strain–displacement matrix $[B]$ as shown in Section 7.7, partial derivatives of the shape function $\phi_i(x)$ are required. Differentiating Eq. (7.46) with respect to the spatial coordinate x_k, we get

$$\phi_{i,k} = \sum_{j=1}^{m} p_{j,k}[[A]^{-1}[C]]_{ji} + p_j[[A]_{,k}^{-1}[C] + [A]^{-1}[C]_{,k}]_{ji} \quad (7.49)$$

where $[C] = [P]^{T}[W]$ and the index following a comma denotes a spatial derivative. To obtain the derivative $[\mathbf{A}]_{,k}^{-1}$, the identity $[A][A]^{-1} = [I]$ is differentiated, where $[I]$ is the identity matrix. Thus,

$$[A][A]_{,k}^{-1} + [A]_{,k}[A]^{-1} = [0] \quad (7.50)$$

leading to

$$[A]_{,k}^{-1} = -[A]^{-1}[A]_{,k}[A]^{-1} \quad (7.51)$$

As seen from Eq. (7.51), no further matrix inversions are required to obtain $[A]_{,k}^{-1}$ as well as the derivative $\phi_{i,k}$ of the shape function.

7.6 MLS WEIGHT FUNCTIONS

The MLS weight functions are introduced into the EFG formulation via the L_2 norm defined in Eq. (7.22), which is minimized to obtain the unknown coefficients $\{a(x)\}$. These weight functions constitute the matrix $[W]$ and are used in defining the EFG shape functions in Eq. (7.46). As stated in Section 7.5, these functions, together with the basis functions $p_j(x)$, determine the differentiability of the EFG shape functions $\phi_i(x)$. The weight function $w_i(x)$, corresponding to a node i, is defined such that it is a monotonically decreasing function of $\|x - x_i\|$. The domain in which the value of the weight function is nonzero is called the *support* of the weight function. From Eq. (7.46) it can be seen that $\phi_i(x) = 0$ when $w_i(x) = 0$. The fact that $\phi_i(x)$ vanishes when x does not lie in the support of nodal point x_i imparts a local character to the MLS approximation.

Several different weight functions have been used in the EFG formulations. Some of the more prominent ones are given below. In the expression for weight functions, $d_i = \|x - x_i\|$ is the distance from node x_i to x, and r_i is the size of the support for the weight function w_i and determines the support at node x_i.

1. Exponential weight function:

$$w_i(\underset{\sim}{x}) = w(r_i) = \begin{cases} \dfrac{\exp[-(d_i/c_i)^{2k_i}] - \exp[-(r_i/c_i)^{2k_i}]}{1 - \exp[-(r_i/c_i)^{2k_i}]} & d_i \le r_i \\ 0 & d_i > r_i \end{cases}$$

(7.52)

where c_i is a constant controlling the shape of the weight function (Belytschko et al., 1994a; Hegen, 1996), and k_i is a constant exponent generally chosen to be unity.

2. Spline weight function:

$$w_i(\underset{\sim}{x}) = w_i(r_i)$$

$$= \begin{cases} 1 - 6\left(\dfrac{d_i}{r_i}\right)^2 + 8\left(\dfrac{d_i}{r_i}\right)^3 - 3\left(\dfrac{d_i}{r_i}\right)^4 & 0 \le d_i \le r \\ 0 & d_i > r_i \end{cases}$$

(7.53)

3. B-spline weight function:

$$w_i(\underset{\sim}{x}) = w(r_i) = \begin{cases} \frac{2}{3} - 4\xi^2 + 4\xi^3 & \xi \le \frac{1}{2} \\ \frac{4}{3} - 4\xi + 4\xi^2 - \frac{4}{3}\xi^3 & \frac{1}{2} \le \xi \le 1 \\ 0 & \xi > 1 \end{cases}$$ (7.54)

where $\xi = d_i/r_i$.

4. Conical weight function:

$$w_i(\underset{\sim}{x}) = w(r_i) = \begin{cases} 1 - (d_i/r_i)^{2k_i} & d_i \le r_i \\ 0 & d_i > r_i \end{cases}$$

(7.55)

where k_i is a real constant generally chosen to be unity.

5. Singular weight function:

$$w_i(\underset{\sim}{x}) = w(r_i) = s_i(\underset{\sim}{x})\|\underset{\sim}{x} - \underset{\sim}{x_i}\|^{-2\alpha_i}$$

(7.56)

where α_i is a real constant and $s_i(\underset{\sim}{x})$ is one of the weight functions defined previously. Generally, $\alpha_i = 1$.

Other weight functions are possible and have been employed in meshless analyses. In defining the weight functions, the size of its support, r_i, must be carefully specified. A necessary condition for an MLS approximation is that at least m (= number of basis functions employed) weight functions are nonzero to ensure the regularity of matrix $[A]$. The support size, r_i, then must be large enough to have a sufficient number of nodes covered in the domain of each nodal point with which the weight function $w_i(x)$ is associated (i.e., $N \geq m$). A large value of r_i will cover a large number of nodes and lead to an increase in the cost of computations. A small size of the support of a weight function is economical but may lead to a non-singular definition of the matrix $[A]$. A smaller support domain also produces a relatively complex shape function, due to the almost singular shape of the weight function.

The support of a weight function at a node is also termed the *domain of influence* of that node in the meshless analysis literature. In the descriptions of weight functions above, a circular domain of influence with radius r_i is assumed. A rectangular domain of influence can also be employed. For rectangular domains, product weights may be employed as $w_i(x) = w(r_x)w(r_y)$, where $r_x = \|x - x_i\|$ and $r_y = \|y - y_i\|$.

7.7 DISCRETE ELEMENT-FREE GALERKIN FORMULATION

The discrete equations describing the behavior of the solid object are obtained using the weak form of the governing equations. Since the shape functions $\phi_i(x)$ do not possess the selectivity property, the trial function u does not satisfy the essential boundary conditions. Hence, one of the weak forms W2, W3, or W4 appropriate for such conditions as described in Section 7.3 must be used. The weak form W2 given in Eq. (7.14) is employed here to demonstrate the procedure.

Following Eq. (7.45), the trial and test functions are approximated using the MLS shape functions as

$$u = \sum_{a=1}^{N} \phi_a(x)u_a \tag{7.57}$$

$$\delta v = \sum_{b=1}^{N} \phi_b(x)\, \delta v_b \tag{7.58}$$

Both the x and y components of test and trial functions are approximated similarly. The Lagrange multiplier λ and its variation $\delta\lambda$ are expressed as

$$\lambda(\underline{x}) = \sum_{a=1}^{k} \psi_a(\xi(\underline{x}))\lambda_a \tag{7.59}$$

$$\delta\lambda(\underline{x}) = \sum_{b=1}^{k} \psi_b(\xi(\underline{x}))\,\delta\lambda_b \tag{7.60}$$

where $\xi(\underline{x})$ is the normalized arc length along the boundary and ψ_a are approximation functions for λ on Γ_u. For example, Lagrange interpolants may be used to represent ψ_a.

A quadratic variation of λ along the boundary may be represented by Lagrange shape functions ψ_i, $i = 1, 2, 3$ as

$$\psi_1(\xi) = -\tfrac{1}{2}\,\xi(1 - \xi) \tag{7.61}$$

$$\psi_2(\xi) = \tfrac{1}{2}\,\xi(1 + \xi) \tag{7.62}$$

$$\psi_3(\xi) = (1 + \xi)(1 - \xi) \tag{7.63}$$

The displacement field, $\underline{u} = \lfloor u_x u_y \rfloor$, consisting of the x- and y-displacement components for a two-dimensional analysis, is written in terms of nodal displacements using Eq. (7.45) as

$$\begin{Bmatrix} u_x \\ u_y \end{Bmatrix} = \begin{bmatrix} \phi_1 & 0 & \phi_2 & 0 & \cdots & \phi_N & 0 \\ 0 & \phi_1 & 0 & \phi_2 & \cdots & 0 & \phi_N \end{bmatrix} \begin{Bmatrix} u_x^1 \\ u_y^1 \\ u_x^2 \\ u_y^2 \\ \vdots \\ u_x^N \\ u_y^N \end{Bmatrix} = [\Phi]\{u\} \tag{7.64}$$

The two-dimensional strain field is next written for small displacement analysis using Eq. (7.3) as

$$\left\{ \begin{array}{c} \varepsilon_x \\ \varepsilon_y \\ \gamma_{xy} \end{array} \right\} = \left\{ \begin{array}{c} \dfrac{\partial u_x}{\partial x} \\[2mm] \dfrac{\partial u_y}{\partial y} \\[2mm] \dfrac{\partial u_x}{\partial y} + \dfrac{\partial u_y}{\partial x} \end{array} \right\} \tag{7.65}$$

where $\gamma_{xy} = 2\varepsilon_{xy}$ is referred to as *engineering shear strain*. Substituting in Eq. (7.65) from Eq. (7.64), the strain field is written as

$$\left\{ \begin{array}{c} \varepsilon_x \\ \varepsilon_y \\ \gamma_{xy} \end{array} \right\} = \left[\begin{array}{cccccc} \phi_{1,x} & 0 & \phi_{2,x} & 0 & \cdots & \phi_{N,x} & 0 \\ 0 & \phi_{1,y} & 0 & \phi_{2,y} & \cdots & 0 & \phi_{N,y} \\ \phi_{1,y} & \phi_{1,x} & \phi_{2,y} & \phi_{2,x} & \cdots & \phi_{N,y} & \phi_{N,x} \end{array} \right] \left\{ \begin{array}{c} u_x^1 \\ u_y^1 \\ u_x^2 \\ u_y^2 \\ \vdots \\ u_x^N \\ u_y^N \end{array} \right\} \tag{7.66}$$

which is expressed in shorthand matrix notation as

$$\{\varepsilon\} = [B]\{u\} \tag{7.67}$$

$$\{\varepsilon\} = \left\{ \begin{array}{c} \varepsilon_x \\ \varepsilon_y \\ \gamma_{xy} \end{array} \right\} \qquad \{u\} = \left\{ \begin{array}{c} u_x^1 \\ u_y^1 \\ u_x^2 \\ u_y^2 \\ \vdots \\ u_x^N \\ u_y^N \end{array} \right\} \tag{7.68}$$

$$[B] = [[B_1][B_2] \cdots [B_N]] \tag{7.69}$$

$$[B_a] = \left[\begin{array}{cc} \phi_{a,x} & 0 \\ 0 & \phi_{a,y} \\ \phi_{a,y} & \phi_{a,x} \end{array} \right] \tag{7.70}$$

The Lagrange multiplier field is expressed in the matrix notation using Eq. (7.59) as

$$\left\{ \begin{array}{c} \lambda_x \\ \lambda_y \end{array} \right\} = \left[\begin{array}{cccccccc} \psi_1 & 0 & \psi_2 & 0 & \cdots & \psi_k & 0 \\ 0 & \psi_1 & 0 & \psi_2 & \cdots & 0 & \psi_k \end{array} \right] \left\{ \begin{array}{c} \lambda_x^1 \\ \lambda_y^1 \\ \lambda_x^2 \\ \lambda_y^2 \\ \vdots \\ \lambda_x^k \\ \lambda_y^k \end{array} \right\} = [N]\{\lambda\} \quad (7.71)$$

where

$$[N] = [[N_1][N_2] \cdots [N_k]] \qquad (7.72)$$

$$[N_b] = \left[\begin{array}{cc} \psi_b & 0 \\ 0 & \psi_b \end{array} \right] \qquad (7.73)$$

$$\{\lambda\} = \left\{ \begin{array}{c} \lambda_x^1 \\ \lambda_y^1 \\ \lambda_x^2 \\ \lambda_y^2 \\ \vdots \\ \lambda_x^k \\ \lambda_y^k \end{array} \right\} \qquad (7.74)$$

In Eq. (7.74), k is the number of boundary nodes on which the essential boundary conditions are prescribed. Similarly, the variational quantities are expressed using the shape function–based descriptions as

$$\{\delta\varepsilon\} = [B]\{\delta v\} \qquad (7.75)$$

$$\left\{ \begin{array}{c} \delta\lambda_x \\ \delta\lambda_y \end{array} \right\} = [N]\{\delta\lambda\} \qquad (7.76)$$

with

$$\{\delta v\} = \left\{ \begin{array}{c} \delta v_x^1 \\ \delta v_y^1 \\ \delta v_x^2 \\ \delta v_y^2 \\ \vdots \\ \delta v_x^N \\ \delta v_y^N \end{array} \right\} \qquad \{\delta\lambda\} = \left\{ \begin{array}{c} \delta\lambda_x^1 \\ \delta\lambda_y^1 \\ \delta\lambda_x^2 \\ \delta\lambda_y^2 \\ \vdots \\ \delta\lambda_x^k \\ \delta\lambda_y^k \end{array} \right\} \qquad (7.77)$$

The stress–strain relation of Eq. (7.4) can be written in the matrix form as

$$\{\sigma\} = [D]\{\varepsilon\} \tag{7.78}$$

where for two-dimensional plane stress analysis,

$$[D] = \frac{E}{1 - \nu^2} \begin{bmatrix} 1 & \nu & 0 \\ \nu & 1 & 0 \\ 0 & 0 & (1 - \nu)/2 \end{bmatrix} \tag{7.79}$$

in which E and ν are the Young's modulus and Poisson's ratio, respectively.

For two dimensions, the weak form W_2 in Eq. (7.14) is now expressed in matrix notation as

$$\int_\Omega \begin{Bmatrix} \delta\varepsilon_x \\ \delta\varepsilon_y \\ \delta\gamma_{xy} \end{Bmatrix}^{\mathrm{T}} \begin{Bmatrix} \sigma_x \\ \sigma_y \\ \sigma_{xy} \end{Bmatrix} d\Omega - \int_\Omega \begin{Bmatrix} \delta v_x \\ \delta v_y \end{Bmatrix}^{\mathrm{T}} \begin{Bmatrix} F_x \\ F_y \end{Bmatrix} d\Omega - \int_{\Gamma_t} \begin{Bmatrix} \delta v_x \\ \delta v_y \end{Bmatrix}^{\mathrm{T}} \begin{Bmatrix} \bar{t}_x \\ \bar{t}_y \end{Bmatrix} d\Gamma$$

$$- \int_{\Gamma_u} \begin{Bmatrix} \delta\lambda_x \\ \delta\lambda_y \end{Bmatrix}^{\mathrm{T}} \begin{Bmatrix} u_x \\ u_y \end{Bmatrix} d\Gamma - \int_{\Gamma_u} \begin{Bmatrix} \delta v_x \\ \delta v_y \end{Bmatrix}^{\mathrm{T}} \begin{Bmatrix} \lambda_x \\ \lambda_y \end{Bmatrix} d\Gamma$$

$$+ \int_{\Gamma_u} \begin{Bmatrix} \delta\lambda_x \\ \delta\lambda_y \end{Bmatrix}^{\mathrm{T}} \begin{Bmatrix} \bar{u}_x \\ \bar{u}_y \end{Bmatrix} d\Gamma = 0 \tag{7.80}$$

A few key terms in Eq. (7.80) are expanded explicitly to show the process followed to obtain the system of EFG matrix equations. The first term in Eq. (7.80) is written as

$$\int_\Omega \begin{Bmatrix} \delta\varepsilon_x \\ \delta\varepsilon_y \\ \delta\gamma_{xy} \end{Bmatrix}^{\mathrm{T}} \begin{Bmatrix} \sigma_x \\ \sigma_y \\ \sigma_{xy} \end{Bmatrix} d\Omega = \int_\Omega \{\delta v\}^{\mathrm{T}} [B]^{\mathrm{T}}[D][B]\{u\} \, d\Omega \tag{7.81}$$

where substitutions have been made from relations in Eqs. (7.75), (7.67), and (7.78) in obtaining the right-hand side of Eq. (7.81). The right-hand side of Eq. (7.81) is next written as

$$\int_\Omega \{\delta v\}^\mathrm{T} [B]^\mathrm{T}[D][B]\{u\} \; d\Omega = \{\delta v\}^\mathrm{T} [K]\{u\} \qquad (7.82)$$

where the stiffness matrix $[K]$ is

$$[K] = \int_\Omega [B]^\mathrm{T}[D][B] \; d\Omega \qquad (7.83)$$

The integrand in Eq. (7.83) is expanded using the relation in Eq. (7.69) as

$$[B]^\mathrm{T} [D][B] = \begin{bmatrix} [B_1]^\mathrm{T} \\ [B_2]^\mathrm{T} \\ \vdots \\ [B_N]^\mathrm{T} \end{bmatrix}_{2N \times 3} [D]_{3 \times 3} \; [[B_1][B_2] \cdots [B_N]]_{3 \times 2N}$$

$$= \begin{bmatrix} [B_1]^\mathrm{T}[D][B_1] & [B_1]^\mathrm{T}[D][B_2] & \cdots & [B_1]^\mathrm{T}[D][B_N] \\ [B_2]^\mathrm{T}[D][B_1] & [B_2]^\mathrm{T}[D][B_2] & \cdots & [B_2]^\mathrm{T}[D][B_N] \\ \vdots & & & \\ [B_N]^\mathrm{T}[D][B_1] & [B_N]^\mathrm{T}[D][B_2] & \cdots & [B_N]^\mathrm{T}[D][B_N] \end{bmatrix}$$

$$(7.84)$$

The stiffness matrix $[K]$ is seen in Eq. (7.84) to be composed of sub-matrices $[K_{ab}]$, where

$$[K_{ab}] = \int_\Omega [B_a]^\mathrm{T} [D][B_b] \; d\Omega$$

$$a = 1, 2, \ldots N; \quad b = 1, 2, \ldots, N \qquad (7.85)$$

The submatrix $[K_{ab}]$ in Eq. (7.85) premultiplies the displacement vector $\{u_b\}$ in Eq. (7.82). The second term in Eq. (7.80) is now expanded as

$$\int_\Omega \begin{Bmatrix} \delta v_x \\ \delta v_y \end{Bmatrix}^\mathrm{T} \begin{Bmatrix} F_x \\ F_y \end{Bmatrix} \; d\Omega = \int_\Omega \{\delta v\}^\mathrm{T} [\Phi]^\mathrm{T} \begin{Bmatrix} F_x \\ F_y \end{Bmatrix} \; d\Omega \qquad (7.86)$$

where the variational displacement field, $\delta v = \lfloor \delta v_x \delta v_y \rfloor$, is expressed similar to the expression for displacement field in Eq. (7.64). The right-hand side of Eq. (7.86) may be written as

$$\int_{\Omega} \{\delta v\}^{\mathrm{T}} [\Phi]^{\mathrm{T}} \begin{Bmatrix} F_x \\ F_y \end{Bmatrix} d\Omega = \{\delta v\}^{\mathrm{T}} \{F\} \qquad (7.87)$$

where the force vector $\{F\}$ is written as

$$\{F\} = \int_{\Omega} [\Phi]^{\mathrm{T}} \begin{Bmatrix} F_x \\ F_y \end{Bmatrix} d\Omega \qquad (7.88)$$

The integrand in Eq. (7.88) is further expanded by substituting for $[\Phi]$ from Eq. (7.64) to obtain

$$[\Phi]^{\mathrm{T}} \begin{Bmatrix} F_x \\ F_y \end{Bmatrix} = \begin{Bmatrix} \phi_1 & 0 \\ 0 & \phi_1 \\ \phi_2 & 0 \\ 0 & \phi_2 \\ \vdots & \vdots \\ \phi_N & 0 \\ 0 & \phi_N \end{Bmatrix} \begin{Bmatrix} F_x \\ F_y \end{Bmatrix} = \begin{Bmatrix} \phi_1 \begin{Bmatrix} F_x \\ F_y \end{Bmatrix} \\ \phi_2 \begin{Bmatrix} F_x \\ F_y \end{Bmatrix} \\ \vdots \\ \phi_N \begin{Bmatrix} F_x \\ F_y \end{Bmatrix} \end{Bmatrix} \qquad (7.89)$$

The force vector $\{F\}$ is seen from Eq. (7.89) to be composed of subvectors $\{f_b^\Omega\}$, where

$$\{f_b^\Omega\} = \int_{\Omega} \phi_b \begin{Bmatrix} F_x \\ F_y \end{Bmatrix} d\Omega \qquad b = 1, 2, \ldots, N \qquad (7.90)$$

The remaining terms in Eq. (7.80) may be expanded similarly to obtain

$$\{\delta v\}^{\mathrm{T}} [[K_{ab}]\{u_b\} - \{f_b^\Omega\} - \{f_b^\Gamma\} + [G_{ab}]\{\lambda_b\}]$$
$$+ \{\delta\lambda\}^{\mathrm{T}} [[G_{ab}]^{\mathrm{T}} \{u^b\} - \{q_b\}] = 0 \qquad (7.91)$$

Since $\{\delta v\}$ and $\{\delta\lambda\}$ are arbitrary variations, we obtain two matrix equations by equating the terms that postmultiply each of them separately to $\{0\}$ as

$$[K_{ab}]\{u_b\} - \{f_b^\Omega\} - \{f_b^\Gamma\} + [G_{ab}]\{\lambda_b\} = \{0\} \qquad (7.92)$$
$$[G_{ab}]^{\mathrm{T}} \{u_b\} - \{q_b\} = \{0\} \qquad (7.93)$$

Writing these equations together in matrix form, we have

$$\begin{bmatrix} [K_{ab}] & [G_{ab}] \\ [G_{ab}]^{\mathrm{T}} & [0] \end{bmatrix} \begin{Bmatrix} \{u_b\} \\ \{\lambda_b\} \end{Bmatrix} = \begin{Bmatrix} \{f_b\} \\ \{q_b\} \end{Bmatrix} \tag{7.94}$$

with

$$\{f_b\} = \{f_b^{\Omega}\} + \{f_b^{\Gamma}\} \tag{7.95}$$

$$[G_{ab}] = -\int_{\Gamma} \phi_a [N_b]\, d\Gamma \tag{7.96}$$

$$\{f_b^{\Gamma}\} = \int_{\Gamma_t} \phi_b \begin{Bmatrix} \bar{t}_x \\ \bar{t}_y \end{Bmatrix} d\Gamma \tag{7.97}$$

$$\{q_b\} = -\int_{\Gamma_u} \psi_b \begin{Bmatrix} \bar{u}_x \\ \bar{u}_y \end{Bmatrix} d\Gamma \tag{7.98}$$

Finally, the contributions of the form of Eq. (7.94) for each nodal point are assembled to obtain a global system of equations as

$$\begin{bmatrix} [K] & [G] \\ [G]^{\mathrm{T}} & [0] \end{bmatrix} \begin{Bmatrix} \{u\} \\ \{\lambda\} \end{Bmatrix} = \begin{Bmatrix} \{f\} \\ \{q\} \end{Bmatrix} \tag{7.99}$$

In Eq. (7.99), $\{u\}$ and $\{\lambda\}$ are the vectors of unknown nodal displacements and Lagrange multipliers, respectively. The system of equations (7.99) may be solved using standard linear system solvers to obtain these unknown quantities.

7.8   NUMERICAL IMPLEMENTATION

The computation of the matrices $[K]$, $[G]$, $\{f\}$ and $\{q\}$ in Eq. (7.99) requires evaluation of various domain and boundary integrals, such as in Eqs. (7.85), (7.90), and (7.95) to (7.98). To evaluate the domain integrals, the domain Ω is discretized into volume cells. Similarly, to evaluate the boundary integrals on portions Γ_u and Γ_t of the boundary, these portions are discretized into boundary cells. The volume and boundary cells for an arbitrary two-dimensional body are shown in Fig. 7.1. Numerical integration (e.g., Gauss quadrature) is then performed over each of these cells to obtain the cell contributions $[K_c]$, $[G_c]$, $\{f_c\}$, and $\{q_c\}$.

The procedure to obtain the cell contributions $[K_c]$, $[G_c]$, $\{f_c\}$, and $\{q_c\}$ is outlined below. A typical volume cell with a 4×4 distribution

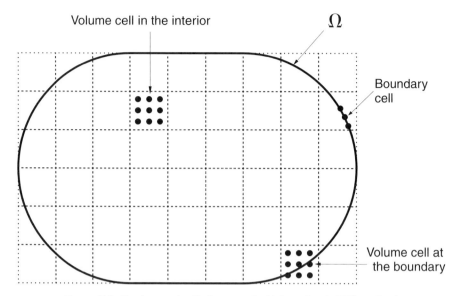

Figure 7.1 Background cells for numerical integration in EFG method.

of Gauss integration points is shown in Fig. 7.2. The computation steps performed at a typical integration (quadrature) point x_q in a cell are as follows:

Step 1. Determine all nodes x_i such that the integration point x_q lies within the support of the weight function at x_i [i.e., $w_i(x_q) > 0$].

Step 2. At each of the node x_i found in step 1, compute the shape functions $\phi_i(x_q)$ and their derivatives $\phi_{i,j}(x_q)$.

Step 3. Compute the contributions of x_q to the integrals in Eqs. (7.85) and (7.90).

Step 4. Assemble these contributions from step 3 into cell contributions $[K_c]$ and $\{f_c\}$, respectively, by suitably multiplying them with Gaussian (or other) weight functions and adding together.

Step 5. Assemble the cell contributions in overall matrices $[K]$ and $\{f\}$, respectively.

Cell integrations for boundary cells are performed using similar steps to obtain contributions $[G_c]$, $\{f_c\}$, and $\{q_c\}$, which are assembled into the overall matrices $[G]$, $\{f\}$, and $\{q\}$.

As stated in Section 7.4, the MLS shape functions are not piecewise polynomials. This requires a relatively dense distribution of quadrature

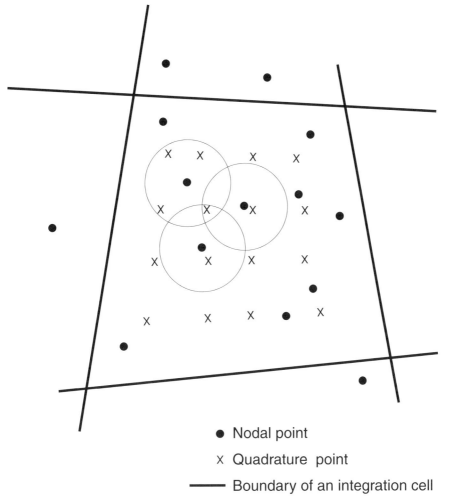

● Nodal point

x Quadrature point

—— Boundary of an integration cell

Figure 7.2   Typical integration cell, nodal points, and quadrature points in EFG method.

points in each cell for accurate evaluation of the respective integrals. Various suggestions for the number of integration points needed for accurate evaluation of integrals have been made in the literature. For example (Hegen, 1996), a total of n_{int} integration points may be used in the domain with

$$n_{int} = \mu N \qquad (7.100)$$

where N is the number of nodal points in the domain and $5 \le \mu \le 8$. For objects with domains that have curved boundaries, a background cell structure will produce cells on the boundary for which the integration points may not lie within the domain of the boundary. One such cell is shown in Fig. 7.1, where for a 3 × 3 Gauss integration distribution of quadrature points, four such points fall outside the domain Ω. In several implementations (Belytschko et al., 1994a; Lu et al., 1994), an integration point is simply skipped when it falls outside the domain of the body. In other implementations (Hegen, 1996), the background cells are constructed using isoparametric mapping such that the cells on the boundary conform to the curved shape of the boundary.

The desire to remove the need for a mesh in finite element analyses has been a major motivating factor for the development of meshless methods. The use of background integration cells as described above thus constitutes a major criticism of the EFG and other meshless methods. The background cells are viewed as a type of mesh similar to the mesh for finite element analysis. Several attempts have been reported in the literature (Beissel and Belytschko, 1996; Chen et al., 2001; Yoo et al., 2001) to develop a node-based integration that does not require the background cells. Nodal integration usually introduces numerical instabilities. Stabilization terms may be added to improve the performance of nodal integration. For problems that do not contain unstable modes in their original solution, the addition of stabilization actually decreases the accuracy. The stabilized conforming nodal integration scheme (Chen et al., 2001) has shown a great deal of promise, but the scheme requires the Vornoi tessellation (O'Rourke, 2000) of a set of nodes. The tessellation does appear like a background mesh, even though it does not perform the same function as the finite element mesh.

7.9   TREATMENT OF BOUNDARY CONDITIONS

As shown in Section 7.4, the MLS shape functions do not satisfy the Kronecker delta condition [i.e., $\phi_i(x_j) \neq \delta_{ij}$], so that

$$u^h(x_p) = \sum_{i=1}^{N} \phi_i(x_p)u_i$$

$$\neq u_p \tag{7.101}$$

If u_p were the prescribed value at a node x_p on Γ_u, the MLS approximant does not reproduce this value at x_p. This causes a serious problem in terms of direct enforcement of essential boundary conditions. A number of techniques are available to circumvent this difficulty. However, each has its limitations and no method to date has received universal acceptance. Some of the techniques available to account for the essential boundary conditions in an EFG formulation are outlined briefly below.

Lagrange Multiplier Approach   This approach employs Lagrange multipliers for the enforcement of essential boundary conditions. The approach was described in detail earlier and results in the system of equations shown in Eq. (7.99). This approach has several limitations: (1) It leads to larger system matrices since additional unknowns corresponding to the Lagrange multipliers are introduced; (2) the bandedness of the system is sacrificed due to the appearance of matrices $[G]$ and $[G]^T$; and (3) the matrix, which has to be inverted, possesses zeros on its main diagonals. Solvers with variable bandwith that do not take advantage of positive definiteness of the matrix need to be employed (Lu et al., 1994).

Penalty Parameter Approach   The essential boundary conditions are incorporated into weak form W4 of the governing equations using a penalty parameter α as shown in Eq. (7.16). An analysis on W4 is performed similar to that performed on weak form W2 in Section 7.6. This results in a set of equations given as

$$[[K] + \alpha[K^u]] \{u\} = \{f\} + \alpha\{f^u\} \qquad (7.102)$$

Additional terms appearing with the penalty parameter are

$$[K_{ab}^u] = \int_{\Gamma_u} \phi_a[S]\phi_j \, d\Gamma \qquad (7.103)$$

$$\{f_b^u\} = \int_{\Gamma_u} \phi_b[S] \begin{Bmatrix} \overline{u}_x \\ \overline{u}_y \end{Bmatrix} d\Gamma \qquad (7.104)$$

where

$$[S] = \begin{bmatrix} S_x & 0 \\ 0 & S_y \end{bmatrix} \qquad (7.105a)$$

$$S_i = \begin{cases} 1 & \text{if } u_i \text{ is prescribed on } \Gamma_u \\ 0 & \text{if } u_i \text{ is not prescribed on } \Gamma_u \end{cases} \qquad (7.105b)$$

A large value of the penalty parameter in the range $\alpha = 10^3$ to 10^7 is employed. The solution of the system of equations may be sensitive to the value of the penalty parameter α.

Transformation Methods   As explained in Section 7.4, the vector $\{u\}$ in Eq. (7.99) contains generalized nodal displacements, not the actual nodal displacements, as a result of which the essential boundary conditions cannot be imposed directly.

A transformation is sought that converts the generalized displacements $\{u\}$ to the actual displacements $\{\hat{u}\}$ at the nodal locations. Requiring that the approximation in Eq. (7.45) yields the actual displacement $\hat{u}_{iJ}$ at location $\underset{\sim}{x}_J$, we obtain

$$\hat{u}_{iJ} = \sum_{I=1}^{N} u_{iI}\phi_I(\underset{\sim}{x}_J) \qquad J = 1, 2, \ldots, N$$

$$= \sum_{I=1}^{N} \Lambda_{IJ}u_{iI} \tag{7.106}$$

where i is the coordinate direction and

$$\Lambda_{IJ} = \phi_I(\underset{\sim}{x}_J) \tag{7.107}$$

In matrix form, we can write

$$\{\hat{u}_i\} = [\Lambda]^{\mathrm{T}}\{u_i\} \tag{7.108a}$$

or

$$\{u_i\} = [\Lambda]^{-\mathrm{T}}\{\hat{u}_i\} \tag{7.108b}$$

In component form,

$$u_{iI} = \sum_{K=1}^{N} \Lambda_{IK}^{-\mathrm{T}}\hat{u}_{iK} \tag{7.109}$$

Substituting this relation back into the approximation in Eq. (7.45) yields

$$u_i^h(\underset{\sim}{x}) = \sum_{I=1}^{N} \phi_I(\underset{\sim}{x}) u_{iI}$$

$$= \sum_{I=1}^{N} \phi_I(\underset{\sim}{x}) \sum_{K=1}^{N} \Lambda_{IK}^{-T} \hat{u}_{iK}$$

$$= \sum_{K=1}^{N} \Psi_K(\underset{\sim}{x}) \hat{u}_{iK} \tag{7.110}$$

where

$$\Psi_K(\underset{\sim}{x}) = \sum_{I=1}^{N} \Lambda_{KI}^{-1} \phi_I(\underset{\sim}{x}) \tag{7.111}$$

Thus, a new interpolation is obtained in Eq. (7.110), expressed in terms of the actual displacements $\{\hat{u}\}$ and modified shape functions $\Psi_K(\underset{\sim}{x})$. This new interpolation does satisfy the Kronecker delta property as follows. From Eq. (7.111), $\Psi_K(\underset{\sim}{x}_j)$ is obtained by substituting $\underset{\sim}{x} = \underset{\sim}{x}_j$ as

$$\Psi_K(\underset{\sim}{x}_j) = \sum_{I=1}^{N} \Lambda_{KI}^{-1} \phi_I(\underset{\sim}{x}_j) \tag{7.112}$$

But from Eq. (7.107), $\phi_I(\underset{\sim}{x}_j) = \Lambda_{IJ}$; thus

$$\Psi_K(\underset{\sim}{x}_j) = \sum_{I=1}^{N} \Lambda_{KI}^{-1} \Lambda_{IJ} = \delta_{KJ} \tag{7.113}$$

which is the selectivity property of the interpolant. The matrices $[K]$ and $\{f\}$ in Eq. (7.99) are now computed using the interpolation functions $\Psi_K(\underset{\sim}{x})$ of Eq. (7.110). Since the essential boundary conditions can now be enforced directly, the matrices $[G]$ and $\{q\}$ in Eq. (7.99) are no longer required.

The transformation method described above requires the inversion of a $N \times N$ matrix $[\Lambda]$. A mixed transformation method may be devised (Chen et al., 2000a) in which only generalized displacements corresponding to the nodes at the essential boundary are transformed to the corresponding actual displacements. The generalized displacements at the remaining nodes are not transformed. This mixed transformation still allows direct enforcement of the essential boundary conditions.

Only a $N_u \times N_u$ matrix now need be inverted, where N_u is the number of nodes on the essential boundary.

Coupled Finite Element–EFG Method Approach To take advantage of the direct enforcement of essential boundary conditions available in the finite element method, several formulations coupling the finite element method with the EFG method have been developed (Belytschko et al., 1995c; Liu et al., 1995, 1996; Hegen, 1996). A layer of finite elements is provided along the entire boundary or at least along that portion of the boundary on which essential boundary conditions need to be enforced. The rest of the body is discretized by a distribution of EFG nodal points as shown in Fig. 7.3. The essential boundary conditions are then imposed as they are in standard finite element analysis.

The MLS formulation becomes a finite element interpolation when the domain of the influence coincides with the element. Hegen (1996) took advantage of this characteristic to formulate the finite element–EFG coupling. Another formulation (Belytschko et al., 1995c) utilizes blending functions, R, to combine the two analysis methods as

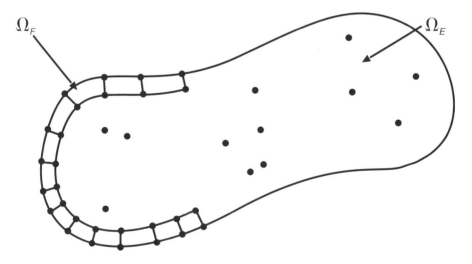

$$\Omega_F \quad \text{Finite element method domain}$$

$$\Omega_E \quad \text{EFG domain}$$

Figure 7.3 Object with coupled finite elements and EFG domains.

$$u = R(\underset{\sim}{x})u^{\text{EFG}}(\underset{\sim}{x}) + [1 - R(\underset{\sim}{x})]u^{\text{FE}}(\underset{\sim}{x}) \qquad (7.114)$$

with the blending function given as

$$R(\underset{\sim}{x}) = \sum_J N_J(\underset{\sim}{x}) \qquad \underset{\sim}{x}_J \in \Gamma_E \qquad (7.115)$$

where Γ_E is the boundary of the portion of the domain modeled using EFG. The introduction of finite elements into a meshless formulation constitutes a major criticism of this approach. It is felt that the meshless character of analysis is compromised by requiring even a portion of the domain to be discretized using finite elements.

Boundary Singular Kernel Method The Kronecker delta property of $\phi_i(\underset{\sim}{x}_j)$ may be achieved by employing singular kernel functions in obtaining the MLS interpolant (Lancaster and Salkausas, 1980). This idea was employed by Kaljevic and Saigal (1997) to obtain a EFG formulation that allows direct imposition of essential boundary conditions. The technique results in satisfying the essential boundary conditions at discrete nodal points on the boundary. The portions of the boundary lying between these nodes, however, do not satisfy the requisite conditions.

Several other methods for enforcing essential boundary conditions in meshless analyses have also been proposed [see Mukherjee and Mukherjee (1997b) and the review by Belytschko et al., (1996)].

7.10 OTHER METHODS FOR MESHLESS ANALYSIS

The essence of the meshless EFG method described in Section 7.9 lies in the approximation of the response $u(x)$ using moving least squares–based nodal shape functions as in Eq. (7.46). Other meshless methods may be devised by arriving at similar node-based shape functions or node-based representations of displacement u. Consider the exact representation of a function obtained from using the sifting property of the Dirac delta function. A function $f(x)$ may be represented as

$$f(x) = \int_{-\infty}^{\infty} \delta(x - s)f(s) \, ds \qquad (7.116)$$

where $\delta(x - s)$ is the Dirac delta function that satisfies the property

$$\int_{-\infty}^{\infty} \delta(x - s) \, ds = 1 \qquad (7.117)$$

A smoothed representation of $f(x)$ may be obtained by replacing the Dirac delta function in Eq. (7.116) by a smooth function, say $\phi(x - s)$, as

$$f_s(x) = \int_{-\infty}^{\infty} \phi(x - s) f(s) \, ds \qquad (7.118)$$

where $f_s(x)$ is a smoothed respresentation of $f(x)$ but is not equal to it. The function $\phi(x - s)$ must satisfy the property satisfied by $\delta(x - s)$ in Eq. (7.117): that

$$\int_{-\infty}^{\infty} \phi(x - s) \, ds = 1 \qquad (7.119)$$

The smooth particle hydrodynamics (SPH) method is one of the earliest particle or meshfree methods. It was developed initially for the study of problems in astrophysics (Lucy, 1977, Gingold and Monaghan, 1977). The method was later extended to apply to problems in solid mechanics (Libersky and Petschek, 1991; Libersky et al., 1993). In the SPH nomenclature the function $\phi(x - s)$ is termed the *kernel function*. Thus, the response $u(x)$ of a solid object is given by using the smoothed representation of Eq. (7.118) as

$$u(x) = \int_{-\infty}^{\infty} \phi(x - s) u(s) \, ds \qquad (7.120)$$

A positive function such as the Gaussian function or the B-spline function is usually employed as the kernel function. In one dimension these functions are given as

1. Gaussian function:

$$\phi(x, s) = g(x, h) = \frac{1}{(\pi h^2)^{n/2}} \exp\left(-\frac{x^2}{h^2}\right) \qquad 1 \le n \le 3 \quad (7.121)$$

where h is a parameter called the *smoothing length*.

2. B-spline function:

$$\phi(x, s) = b(x, h) = \frac{2}{3h} \begin{cases} 1 - \frac{3}{2}q^2 + \frac{3}{4}q^3 & q \le 1 \\ \frac{1}{4}(2 - q)^3 & 1 \le q \le 2 \\ 0 & q \ge 2 \end{cases}$$

(7.122)

where $q = \|x - s\|/h$, h is the smoothing length and the radius of the support of this kernel is $2h$.

A major difficulty in SPH methods arises from the fact that even when the kernel function satisfies equation (7.48), its discrete counterpart does not, so that

$$\sum_I \phi(x - s_I) \Delta s_I \ne 1$$

(7.123)

In the discrete form, therefore, the kernel function does not satisfy the partition of unity condition. Referring to the discussion following Eq. (7.48), the kernel function violates the completeness condition and is not able to represent constant-valued fields. Completeness can, however, be restored in a kernel approximation through a correction transformation. A correction for the constant reproducing condition was given by Shepard (1968). Completeness corrections may also be applied to the derivatives of the approximating kernel functions. Several corrections have been proposed in the literature to provide for derivative reproducing conditions. Prominent among these corrections are those of Monaghan (1988, 1992), Johnson and Beissel (1996), Randles and Libersky (1996), and Krongauz and Belytschko (1997, 1998).

The SPH method suffers from several drawbacks. The motion of the particles becomes unstable under certain tensile stress states. Several strategies to circumvent this limitation have been proposed (e.g., Dyka and Ingel, 1995; Dyka et al., 1995; Morris, 1996; Randles and Libersky, 1996). The SPH formulation leads to undesirable zero-energy modes that pollute the solution. A stress point approach (Vignjevic et al., 2000) has been proposed for the elimination of such modes. Finally, the SPH method does not allow direct imposition of essential boundary conditions. Several remedies to this difficulty have also been proposed (Takeda et al., 1994; Randles and Libersky, 1996; Morris et al., 1997).

Another meshless method is the *reproducing kernel particle method* (RKPM). Liu and Chen (1995) noted that the smoothed representation of equation (7.120) employed by the SPH method leads to several difficulties. When applied to discretized domains, it causes amplitude and phase errors in addition to the deterioration of solution at domain boundaries. They proposed the addition of a correction factor, C, to improve the smoothed representation as

$$u(x) = \int_{-\infty}^{\infty} C(x, x - s)\phi(x - s)u(s) \, ds \qquad (7.124)$$

The correction factor C may be given as

$$C(x, x - s) = \sum_{j=0}^{m} a_j(x)p_j(x, s) \qquad (7.125)$$

where $a_j(x)$ are the unknown coefficients and $p_j(x, s)$ are the basis functions. If the basis functions are chosen to be polynomials, the RKPM reduces to the EFG method described earlier. The RKPM method has been applied to a number of linear and nonlinear analysis problems which are reviewed extensively in Li and Liu (2001).

A recent meshless method is the natural neighbor Galerkin method (Sukumar et al., 1998, 2001). The method requires a Vornoi tessellation (O'Rourke, 2000) of the domain of the solid object. As with other meshless methods, the interpolation scheme is expressed as

$$u^h(x) = \sum_{i=1}^{N} \phi_i(x)u_i \qquad (7.126)$$

where $\phi_i(x)$ are the nodal shape functions. These shape functions are computed based on the concept of natural neighbors and the corresponding interpolation introduced by Sibson (1980). An improvement in this interpolation employs the non-Sibsonian interpolation due to Belikov et al. (1997). The advantages of these natural neighbors–based interpolations include: satisfaction of the partition of unity condition, allowing the constant reproducing conditions; and satisfaction of the selectivity property, allowing direct imposition of essential boundary conditions. The major apparent disadvantage lies in the fact that a Delaunay triangulation of the object is required to obtain the Vornoi tessellation on which the interpolation functions are based. This triangulation is viewed as being similar to the finite element mesh, thus diminishing the meshless character of the method.

A meshless method quite similar to the EFG method but with an improved local (node-based) character is the meshless local Petrov–Galerkin (MLPG) method (Atluri and Zhu, 1998, 2000b). Recall that in the weak formulations, the same interpolation of Eq. (7.48) is employed for both the trial and test functions. In the MLPG formulation, the same trial functions are employed, while the weight functions of the MLS approximation with compact local support are used as the test functions. This imparts a strong local character to the formulation, requiring integration at each nodal point over only the support of that nodal point.

7.11 CLOSING REMARKS

Starting in the mid-1990s, there has been lot of activity toward the development of meshless methods of analysis. Several significant advances have been made and have been reviewed by Belytschko et al. (1996) and Li and Liu (2001). Meshless methods offer the possibility of considerable simplification of the analysis process by eliminating the need for a structured mesh. Among several others, two application problems that have been envisioned to see significant simplification due to the success of meshless methods are (1) the arbitrary propagation of cracks in solid bodies, and (2) solid bodies undergoing severe deformations; such as in the case of metal forming. Indeed, several of the earlier developments concentrated on demonstrating the effectiveness of mesh-free methods in crack propagation (Belytschko et al., 1995a, b). Their success in modeling large deformation and other nonlinear problems has also been amply demonstrated (Chen et al., 1996, 1997, 1998). Several factors, however, continue to hinder the widespread acceptance of these methods for analysis. These include (1) the lack of a truly meshless method—the use of background cells or Delaunay triangulation to discretize the domain reduces the attractiveness of these methods, and (2) the difficulties encountered in the imposition of essential boundary conditions.

Although it was believed initially that meshless methods do not suffer from locking problems, some recent studies have indeed discovered such problems with these methods (Kanok-Nukulchai et al., 2001). Several recent developments, such as the natural neighbors method (Moran and Yoo, 2001) offer the hope of improved performance of these methods. A large number of impressive applications, including sensitivity analysis and shape optimization (Kim et al., 2001), stochastic analysis (Rahman and Rao, 2001), analysis of geologic materials (Belytschko et al., 2000), dynamic fracture (Belytschko and Tabarra, 1996), and multiple-length-scale problems (Chen et al., 2000b), have been reported in the literature. The entire field of meshless analysis, however, is in its infancy. It will require several years of continued development to bring these methods to the maturity of the finite element method, to verify the claims of efficiency and ease made on behalf of these methods, and to realize the promised potential of these methods.

Meshless analysis methods are extremely computationally burdensome, requiring 3 to 10 times the computational resources of the finite element method. No guidelines exist at present to suggest a desirable distribution of nodal points in the domain. Several analysts have, in fact, created background finite element meshes to provide nodal distributions for their meshless analysis. Crack propagation problems, although demonstrated successfully, have required an unusually dense distribution of nodes to allow crack propagation. While this situation could be improved significantly by adaptively providing more nodes near the current location of a crack tip, systematic formulations for such adaptive analyses are not yet available. From the point of view of elimination of the tedium due to the complexity of meshing, meshless methods are required especially for three-dimensional analyses. However, the success of these methods has been demonstrated most often for two-dimensional cases, and only a few three-dimensional studies have been reported (Barry and Saigal, 1999). The analysis of a complex three-dimensional industrial component has not yet been attempted. Meshless methods will gain popularity as their various limitations are addressed and as they begin to be used for practical industrial problems.

REFERENCES

Atluri, S. N., and Zhu, T. 1998. A new meshless local Petrov–Galerkin (MLPG) approach in computational mechanics, *Comput. Mech.*, **22,** 117–127.

Atluri, S. N., and Zhu, T. 2000a. New concepts in meshless methods, *Int. J. Numer. Methods Eng.*, **47,** 537–556.

Atluri, S. N., and Zhu, T.-L. 2000b. The meshless local Petrov–Galerkin (MLPG) approach for solving problems in elastostatics, *Comput. Mech.*, **25,** 169–179.

Atluri, S. N., Sladek, J., Sladek, V., and Zhu, T. 2000. The local boundary integral equation (LBIE) and its meshless implementation for linear elasticity, *Comput. Mech.,* **25,** 180–198.

Atluri, S. N., and Shen, S. 2002. *The Meshless (MLPG) Method,* Tech Science Press, Encino, CA.

Barry, W., and Saigal, S. 1999. A three-dimensional element-free Galerkin elastic and elastoplastic formulation. *Int. J. Numer. Methods Eng.,* **46,** 671–693.

Beissel, S., and Belytschko, T. 1996. Nodal integration of the element-free Galerkin method, *Comput. Methods Appl. Mech. Eng.,* **139,** 49–74.

Belikov, V., Ivanov, V., Kontorovich, V., Korytnok, S., and Semenov, A. 1997. The non-Gibsonian interpolation: a new method of interpolation of the values of a function on an arbitrary set of points, *Comput. Math. Math. Phys.,* **37**(1), 9–15.

Belytschko, T., and Tabbara, M. 1996. Dynamic fracture using element-free Galerkin methods. *J. Comput. Appl. Math.,* **39,** 923–938.

Belytschko, T., Lu, Y. Y., and Gu, L. 1994a. Element Free Galerkin Methods, *Int. J. Numer. Methods Eng.,* **37,** 229–256.

Belytschko, T., Lu, Y. Y., and Gu, L. 1994b. Fracture and crack growth by element-free Galerkin methods, *Model. Simul. Sci. Comput. Eng.,* **2,** 519–534.

Belytschko, T., Lu, Y. Y., and Gu, L. 1995a. Crack propagation by element-free Galerkin methods, *Eng. Fracture Mech.,* **51,** 295–315.

Belytschko, T., Lu, Y. Y., and Gu, L. 1995b. Element-free Galerkin methods for static and dynamic fracture, *Int. J. Solids Struct.,* **32,** 2547–2570.

Belytschko, T., Organ, D., and Krongauz, Y. 1995c. A coupled finite element–element-free Galerkin method, *Comput. Mech.,* **17,** 186–195.

Belytschko, T., Kronagauz, Y., Organ, D., Fleming, M., and Krysl, P. 1996. Meshless methods: an overview and recent developments, *Comput. Methods Appl. Mech. Eng.,* **129,** 3–47.

Belytschko, T., Organ, D., and Gerlach, C. 2000. Element-free Galerkin methods for dynamic fracture in concrete, *Comput. Methods Appl. Mech. Eng.,* **187,** 385–399.

Chati, M. K., and Mukherjee, S. 2000. The boundary node method for three-dimensional problems in potential theory, *Int. J. Numer. Methods Eng.,* **47,** 1523–1547.

Chati, M. K., Mukherjee, S., and Mukherjee, Y. X. 1999. The boundary node method for three-dimensional linear elasticity, *Int. J. Numer. Methods Eng.,* **46,** 1163–1184.

Chen, J. S., Pan, C., Wu, C. T., and Roque, C. 1996. Reproducing kernel particle methods for large deformation analysis of nonlinear structures, *Comput. Methods Appl. Mech. Eng.,* **139,** 195–227.

Chen, J. S., Pan, C., and Wu, C. T. 1997. Large deformation analysis of rubber based on a reproducing kernel particle method, *Comput. Mech.*, **19**, 153–168.

Chen, J. S., Roque, C., Pan, C., and Button, S. T. 1998. Analysis of metal forming process based on meshless method, *J. Mater. Process. Technol.*, **80–81**, 642–646.

Chen, J. S., Wang, H. P., Yoon, S., and You, Y. 2000a. Some recent improvement in meshfree methods for incompressible finite elasticity boundary value problems with contact, *Computat. Mech.*, **25**, 137–156.

Chen, J. S., Wu, C. T., and Belytschko, T. 2000b. Regularization of material instabilities by meshfree approximations with intrinsic length scales, *Int. J. Numer. Methods Eng.*, **47**, 1303–1322.

Chen, J. S., Wu, C. T., Yoon, S., and You, Y. 2001. A stabilized conforming nodal integration for Galerkin meshfree methods, *Int. J. Numer. Methods Eng.*, **50**, 435–466.

Chong, K. P., Saigal, S., Thynell, S., and Morgan, H. S. (eds.) 2002. *Modeling and Simulation-based Life Cycle Engineering,* Spon Press, New York.

De, S., and Bathe, K. J. 2000. The method of finite spheres. *Comput. Mech.*, **25**, 329–345.

De, S., and Bathe, K. J. 2001. Displacement/pressure mixed interpolation in the method of finite spheres, *Int. J. Numer. Methods Eng.*, **51**, 275–292.

Duarte, C. A., and Oden, J. T. 1996. H-p clouds: an h-p meshless method, *Numer. Methods Partial Differential Equations,* **12**, 673–705.

Dyka, C. T., and Ingel, R. P. 1995. An approach for tension instability in smoothed particle hydrodynamics, *Comput. Struct.*, **57**, 573–580.

Dyka, C. T., Randles, P. W., and Ingel, R. P. 1995. Stress points for tensor instability in SPH, *Int. J. Numer. Methods Eng.*, **40**, 2325–234.

Gingold, R. A., and Monaghan, J. J. 1977. Smoothed particle hydrodynamics: theory and application to non-spherical stars, *Mon. Not. R. Astron. Soc.*, **181**, 375–389.

Hegen, D. 1996. Element-free Galerkin methods in combination with finite element approaches, *Comput. Methods Appl. Mech. Eng.*, **135**, 143–166.

Johnson, G. R., and Beissel, S. R. 1996. Normalized smoothing functions for SPH impact computations, *Int. J. Numer. Methods Eng.*, **39**, 2725–2741.

Kaljevic, I., and Saigal, S. 1997. An improved element free Galerkin formulation, *Int. J. Numer. Methods Eng.*, **40**, 2953–2974.

Kanok-Nukulchai, W., Barry, W., Saran-Yasoontorn, K., and Bouillard, P. H. 2001. On elimination of shear locking in the element-free Galerkin method, *Int. J. Numer. Methods Eng.*, **52**, 705–725.

Kim, N. H., Choi, K. K., and Chen, J. S. 2001. Die shape design optimization of sheet metal stamping process using meshfree method, *Int. J. Numer. Methods Eng.*, **51**(12), 1385–1405.

Krongauz, Y., and Belytschko, T. 1997. Consistent pseudo-derivatives in meshless methods, *Comput. Methods Appl. Mech. Eng.,* **146,** 371–386.

Krongauz, Y., and Belytschko, T. 1998. EFG approximation with discontinuous derivatives, *Int. J. Numer. Methods Eng.,* **41,** 1215–1233.

Lancaster, P., and Salkauskas, K. 1980. Surface generated by moving least square methods, *Math. Comput.,* **37,** 141–158.

Li, S., and Liu, W. K. 2001. Meshfree and particle methods and their applications, *http://www.ce.berkeley.edu/~shaofan/review.pdf.*

Libersky, L. D. and Petschek, A. G. 1991. Smooth particle hydrodynamics with strength of materials, in *Advances in the Free-Lagrange Method,* Springer-Verlag, New York, pp, 248–257.

Libersky, L. D., Petschek, A. G., Carney, T. C., Hipp, J. R., and Allahdadi, F. A. 1993. High strain Lagrangian in hydrodynamics: a three-dimensional SPH code for dynamic material response, *J. Comput. Phys.,* **109,** 67–75.

Liu, W. K. 1995. An introduction to wavelet reproducing kernel particle methods, *USACM Bull.,* **8,** 3–16.

Liu, W. K., and Chen, Y. 1995. Wavelet and multiple scale reproducing kernel method, *Int. J. Numer. Methods Fluids,* **21,** 901–933.

Liu, W. K., Chen, Y., and Uras, R. A. 1995. Enrichment of the finite element method with the reproducing kernel particle method, in *Current Topics in Computational Mechanics,* Vol. 305, Cory, J. J. F., and Gordon, J. L. (eds.), ASME-PVP, pp. 253–258.

Liu, W. K., Chen, Y., Chang, C. T., and Belytschko, T. 1996. Advances in multiple scale kernel particle methods, *Comput. Mech.,* **18,** 73–111.

Lu, Y. Y., Belytschko, T., and Gu, L. 1994. A new implementation of the element free Galerkin method, *Comput. Methods Appl. Mech. Eng.,* **113,** 397–414.

Lucy, L. B. 1977. A numerical approach to the testing of the fission hypothesis, *Astrophys. J.,* **82,** 1013.

Melenk, J. M., and Babuska, I. 1996. The partition of unity finite element method: basic theory and application, *Comput. Methods Appl. Mech. Eng.,* **139,** 289–314.

Monaghan, J. J. 1988. An introduction to SPH, *Comput. Phys. Commun.,* **48,** 89–96.

Monaghan, J. J. 1992. Smoothed particle hydrodynamics, *Annu. Rev. Astron. Astrophys.,* **30,** 543–574.

Moran, B., and Yoo, J. 2001. Meshless methods for life-cycle engineering simulation: natural neighbor methods, in *Modeling and Simulation-Based Life-Cycle Engineering,* Chong, K. P., Morgan, H. S., and Saigal, S. (eds.), E&FN Spon, London.

Morris, J. P. 1996. Stability properties of SPH, *Publ. Astron. Soc. Aust.,* **13,** 97.

Morris, J. P., Fox, P. J., and Zhu, Y. 1997. Modeling low Reynolds number incompressible flows using SPH, *J. Comput. Phys.,* **136,** 214–226.

Mukherjee, Y. X., and Mukherjee, S. 1997a. The boundary node method in potential problems, *Int. J. Numer. Methods Eng.,* **40,** 797–815.

Mukherjee, Y. X., and Mukherjee, S. 1997b. On boundary conditions in the element-free Galerkin method, *Comput. Mech.,* **19,** 264–270.

Mukherjee, Y. X., Mukherjee, S., Shi, X., and Nagarajan, A. 1997. The boundary contour method for three-dimensional linear elasticity with a new quadratic boundary element, *Eng. Anal. Boundary Elements,* **20,** 35–44.

Nagarajan, A., Lutz, E., and Mukherjee, S. 1994. A novel boundary element method for linear elasticity with no numerical integration for two-dimensional and line integrals for three-dimensional problems, *ASME J. Appl. Mech.,* **61,** 264–269.

Nagaraj an, A., Mukherjee, S., and Lutz, E. 1996. The boundary contour method for three-dimensional linear elasticity, *ASME J. Appl. Mech.,* **63,** 278–286.

Nayroles, B., Touzot, G., and Villon, P. 1992. Generalization of the finite element method: diffuse approximation and diffuse elements, *Comput. Mech.,* **10,** 307–318.

O'Rourke, J. 2000. *Computational Geometry in C,* Cambridge University Press, Cambridge.

Rahman, S., and Rao, B. N. 2001. A perturbation method for stochastic meshless analysis in elastostatics, *Int. J. Numer. Methods Eng.,* **50,** 1969–1992.

Randles, P. W., and Libersky, L. D. 1996. Smoothed particle hydrodynamics: some recent improvements and applications, *Comput. Methods Appl. Mech. Eng.,* **139,** 375–408.

Schey, H. M. 1973. *Div, Curl, and All That,* W. W. Norton, New York.

Shepard, D. 1968. A two-dimensional interpolation function for irregularly spaced points, in *Proc. ACM National Conference,* pp. 517–524.

Sibson, R. 1980. A vector identity for the Dirichlet tessellation, *Math. Proc. Cambridge Philos. Soc.,* **87,** 151–155.

Sukumar, N., Moran, B., and Belytschko, T. B. 1998. The natural element method in solid mechanics, *Int. J. Numer. Methods Eng.,* **43,** 839–887.

Sukumar, N., Moran, B., Semenov, Y., and Belikov, B. 2001. Natural neighbour Galerkin methods, *Int. J. Numer. Methods Eng.,* **50,** 1–27.

Swegle, J. W., Attaway, S. W., Heinstein, M. W., Mello, F. J., and Hicks, L. 1994. An analysis of smoothed particle hydrodynamics, *Sandia Report SAND-93-2513,* Sandia National Laboratories, Albuquerque, NM.

Takeda, H., Miyama, S. M., and Sekiya, M. 1994. Numerical simulation of viscous flow by smoothed particle hydrodynamics, *Prog. Theor. Phys.,* **116,** 123–134.

Vignjevic, R., Campbell, J., and Libersky, L. 2000. A treatment of zero-energy modes in the smoothed particle hydrodynamics method, *Comput. Methods Appl. Mech. Eng.,* **184,** 67–85.

Yoo, J., Moran, B., and Chen, J. S. 2001. A nodal natural neighbour method, submitted for publication.

Zhu, T., and Atluri, S. N. 1998. A modified collocation method and a penalty formulation for enforcing the essential boundary conditions in the element free Galerkin method, *Comput. Mech.,* **21,** 211–222.

Author Index

259

Subject Index